BLACKWELL HABITAT FIELD GUIDES
Common Plants of Woodlands

Michael Quigley

*Senior Lecturer in Environmental Biology,
Nene College, Northampton*

Norman Copland

*Senior Lecturer in Environmental Biology,
Nene College, Northampton*

First published 1987

Published by Basil Blackwell Ltd
108 Cowley Road
Oxford OX4 1JF
England

British Library Cataloguing in Publication Data

Quigley, Michael
    Common plants of woodlands.—(Blackwell
    habitat field guides).
    1. Forest flora—Great Britain
    I. Title   II. Copland, Norman
    581.941   QK306
    ISBN 0–631–15856–1
    ISBN 0–631–90164–7 School ed.

Cover photograph: Graham Topping

Typeset in 10 on 11pt Garamond
by Opus, Oxford

Printed in Hong Kong

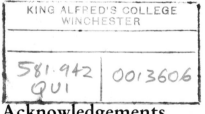
Acknowledgements

We wish to express our thanks to all our colleagues at Nene College who have
supported this venture. Our special thanks go to Dr Eric Ogilvie, Director of
Nene College, who has shown a keen interest in the success of the habitat guides.
Our wives, Roz and Betty, we thank for their patience and encouragement during
the preparation of this guide.

An Introduction to the Habitat Keys

The series of selected habitat keys for flowering and some non-flowering plants has been designed for the beginner in field studies and any other interested person. It is hoped that the keys will provide a straightforward method of identifying the more common plants which occur in the habitats considered.

There is inevitably a flaw in this approach that the user should bear in mind. Since only the more commonly occurring plants are included, it is possible that you will encounter plants which are not mentioned in the keys. It is important that you do not attempt to make a particular plant fit one of the descriptions given. In this woodland plants key, ground cover plants such as mosses and liverworts are not included. Should you wish to identify plants not considered in the keys, it is suggested that you refer to the bibliography for suitable references. For most purposes *The Wild Flower Key* by Francis Rose and *Grasses* by C.E. Hubbard are highly recommended. The key follows the taxonomic treatment and nomenclature of *Flora Europaea*.

How to use the keys

Before you attempt to identify a plant, familiarise yourself with the lay-out of the keys. Scientific and common names are given for all the plants considered in each key. Line diagrams are also provided in the hope that they will support the written descriptions. Do not rely on the diagrams alone to identify the plant. Use the glossary provided to obtain descriptions of terms used in the keys which are unfamiliar to you. You will notice that the book has been divided into sections. In some of these sections, before the actual key begins, numbered descriptions lead you to the appropriate key. When you reach the actual key, the instructions given below should help with your identification.

Always work through the key in a systematic manner using the instructions given below. Do not attempt to make a plant fit a particular description. The plant in question may not be in the keys.

1 On the left-hand side of each page are numbers arranged in sequence with numbers in brackets beside them. The numbers in brackets will enable you to retrace your identification if you feel that you have made a mistake. Each number is set against a pair of contrasting statements **a** and **b**. (In some cases, three statements **a**, **b** and **c** are given.) One of these statements should provide a partial description of the plant which you wish to identify. Examine each plant carefully, using a 10x hand lens where necessary to provide details of the structure. Then study each statement and decide which most closely describes your specimen.
2 Each statement ends *either* in an arrow pointing to a number *or* in a box naming a plant. If the statement ends in an arrow, this means 'go on to' the number indicated. Find this number on the left-hand side of a page and continue your identification until a statement ends by naming a plant. At this point the identification is complete.

Glossary

achene	a small, nut-like fruit
acute	a sharp angle, less than a right-angle
alternate	leaves alternating up the stem — first on one side and then on the other
anther	see **flower**
awn	a stiff, bristle-like projection arising from the spikelets of grasses
axis	main stem running through an inflorescence
basal	leaves at the base of the stem at ground level
bract	a leaf with a flower in its axil
bracteole	a tiny leaf on a flower-stem without a flower in its axil
bristle-like (leaves)	tightly rolled and appearing like a bristle
calyx	see **flower**
capsule	dry fruit that opens into two or more parts or by a lid or holes to release seeds
catkin	a spike of minute flowers, male and female borne separately
compound	a leaf divided into distinct, separate leaflets

trifoliate   palmate   1-pinnate   2-pinnate   3-pinnate

converging	tendency to meet at a point
corolla	see **flower**
corymb	an inflorescence with the outer flower-stalks much longer than the inner ones. The flowers are at roughly the same level in a flat-topped cluster
cyme	an inflorescence in which the top flowers open first. Lower flowers open in sequence, lowest opening last
deciduous	a woody plant which loses its leaves in the autumn
disc	see **florets**
diverging	tendency to spread from a point

alternate

awn

bract

catkin

corymb

cyme

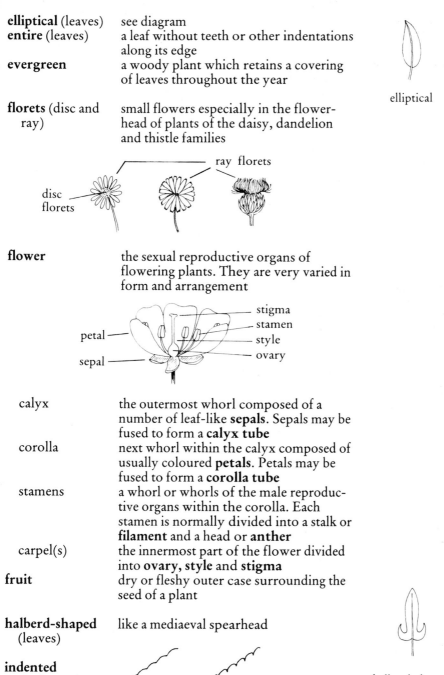

elliptical (leaves)   see diagram

entire (leaves)   a leaf without teeth or other indentations along its edge

evergreen   a woody plant which retains a covering of leaves throughout the year

elliptical

florets (disc and ray)   small flowers especially in the flower-head of plants of the daisy, dandelion and thistle families

ray florets

disc florets

flower   the sexual reproductive organs of flowering plants. They are very varied in form and arrangement

petal — stigma — stamen — style — ovary

sepal

calyx   the outermost whorl composed of a number of leaf-like **sepals**. Sepals may be fused to form a **calyx tube**

corolla   next whorl within the calyx composed of usually coloured **petals**. Petals may be fused to form a **corolla tube**

stamens   a whorl or whorls of the male reproductive organs within the corolla. Each stamen is normally divided into a stalk or **filament** and a head or **anther**

carpel(s)   the innermost part of the flower divided into **ovary**, **style** and **stigma**

fruit   dry or fleshy outer case surrounding the seed of a plant

halberd-shaped (leaves)   like a mediaeval spearhead

indented (leaf margin)

halberd-shaped

inflated   distended, swollen

inflorescence	flowering branch. Any grouping of flowers on a stem or in leaf-axils
lanceolate (leaves)	lance-shaped
leaflets	the separate leaf-blades of a compound leaf
ligule(s) (in grasses)	a small flap of tissue or fringe of hairs where the leaf-blade joins its sheathing base
linear (leaves)	long, narrow, more or less parallel-sided leaf
lyre-shaped (leaves)	see diagram

lanceolate

lyre-shaped

ligule   leaf-blade

membranous (ligules)	a flap of tissue
midrib	the main, central vein of a leaf
oblong (leaves)	a leaf about two or three times as long as broad, parallel-sided in the central part
opposite (leaves)	leaves arising in pairs
oval, ovate (leaves)	rather egg-shaped, about twice as long as broad

oblong   oval

opposite

palmate	see **compound**
panicle	a branched raceme (see **raceme**)
pendulous	hanging down
pinnate (1, 2 or 3)	see **compound**
pinnatifid (leaves)	a deeply cut leaf but not cut right to midrib
prickles	a sharp, usually curved, outgrowth from the outer layers
prostrate	lying in a horizontal position

pinnatifid

raceme	a more or less elongated inflorescence in which the lowest flower opens first, and then the others open in sequence towards the tip
rhizome	a creeping, underground stem
root leaves	leaves arranged at the stem base at ground level

raceme   panicle

4

rosette (of leaves)	leaves arranged in a more or less flat position to the ground
runner	a creeping stem above the ground, can root at tip to form a new plant

rosette

sepals	see **flower**
sessile	without a stalk
sheath	lower part of the leaf surrounding a stem
shrub	a woody plant without a main trunk, branched from the base
simple (leaves)	not divided into leaflets, margin may be entire, lobed or toothed

simple leaf   spike

spike	an unbranched flower-head
spikelet	unit of a grass flower-head
spike-like	resembling a spike
spine	a stiff, straight, sharp-pointed structure
spur (in flowers)	a cylindrical or conical, sometimes curved, hollow projection from the back of certain flowers
stamens	see **flower**
stigma	see **flower**
stipules	leaf-like or scale-like structures at the base of the leaf-stalk or stem
stolon	a creeping stem above ground, not necessarily rooting at tip
strap-shaped	flat, parallel-sided, blunt-tipped leaf

spikelets

stipules

terminal	at the tip
thorn	a woody, sharp-pointed structure
tree	a woody plant with a main trunk
trifoliate	see **compound**
tufted	loose, compact or dense cluster
tussock	a clump

umbel	a flat-topped inflorescence with several branches all arising from one point at the top of the main stem, may be simple or compound
umbel-like	like an umbel inflorescence

umbel

whorl	more than two structures (e.g. leaves) of the same kind arising at the same level

whorl

The Common Plants of the Woodlands

Use the information given below to determine the group to which your plant belongs and proceed to the section indicated by one of the letters A, B, C, D, E or F. (Use the glossary if you do not understand any of the terms used in the key.)

1　Ferns which do not produce flowers bearing sepals, petals, stamens and carpels. Many have finely divided (two-pinnate or three-pinnate), feathery leaves (or fronds). Some have undivided leaves. Spores are borne in spore cases (sori) on the backs of the leaves. Caution: Do not confuse ferns with flowering plants with pinnate leaves (see pages 57–59).

> ### SECTION A,
> page 7

2　Woody or large herbaceous plants which climb, twine, sprawl or scramble over other plants or various available surfaces.

> ### SECTION B,
> page 9

3　Woody shrubs or trees.

> ### SECTION C,
> page 13

4　Herbaceous plants with simple, long, narrow-linear, grass-like leaves with entire margins. Flowers small, usually clustered, individually inconspicuous.

> ### SECTION D,
> page 25

5　Herbaceous plants with simple leaves. Either plant of prostrate habit (stems running along the ground and ascending at the tips) or plant of erect habit.

> ### SECTION E,
> page 31

6　Herbaceous plants with leaves deeply cut or lobed but not to midrib of leaf (pinnatifid) or leaves divided into distinct leaflets (compound).

> ### SECTION F,
> page 49

SECTION A

Ferns

1   **a** Leaves divided into **pinnately arranged leaflets** (pinnae), may appear feathery. ————————————————————————→ **2**
Note that, if leaflets (pinnae) are further divided, the structures are called pinnules.

b Leaves **undivided** and **strap-like**, margin entire or slightly lobed, 10–60 cm long, 30–60 mm wide, arising from the base of the plant. Spore cases arranged in lines, occupying more than half the leaf width, along the backs of the leaves.

part of back of leaf showing spore cases

> *Phyllitis scolopendrium*
> Harts Tongue Fern

2 **(1) a** Leaves **one-pinnate**, lanceolate in outline.

———————————————————————→ **3**

b Leaves **two-pinnate** or **three-pinnate**, appearing feathery.

———————————————————————→ **4**

3 **(2) a** Leaves 5–45 cm long, rather leathery, leaflets (pinnae) fairly blunt, mostly more or less equal in length but tapering towards the top. Pinnae **angled upwards** except the first pair. Spore cases rounded, **in pairs** along pinnae on either side of midrib.

spore cases

part of leaf

> *Polypodium vulgare*
> Common Polypody

b Leaves 10–70 cm long (sterile outer leaves up to 75 cm long and narrowly lanceolate). Pinnae **bluntly oblong** (very narrow on fertile, inner leaves). Spore cases when mature **appearing to cover the whole under-surface** of the pinnae. Fern distinctly tufted.

part of leaf

> *Blechnum spicant*
> Hard Fern

4 (2) a Leaves **two-pinnate**.

                                                    ➤ **5**

   b Leaves **three-pinnate**.

                                                    ➤ **6**

5 (4) a Leaves 30–130 cm long, tapering at both ends. Pinnules (divisions of pinnae) **blunt-tipped**, usually lobed (lobes more or less triangular). Spore cases usually **five** or **six** on each pinnule. Stalks of leaves with pale brown scales.

> *Dryopteris filix-mas*
> Male Fern

spore cases
on pinnule

b Similar in appearance to the above but more graceful. Leaves paler green, up to 150 cm long. Pinnules **pointed tipped**, more toothed. Spore cases often **four-paired**, curved. In damper woods, especially on acid soils.

> *Athyrium filix-femina*
> Lady Fern

spore cases on
pinnule

6 (4) a Leaves, 30–180 cm or more in length, first appear tightly curled (like a bishops-crook) and develop on long stalks up to 4 m in height. Spore cases **covered by inrolled leaf margins**. Fern persists in autumn and winter as copper-brown leaves in a dead state. Mainly on light acid soils.

> *Pteridium aquilinum*
> Bracken

spore cases
at margins

b Leaves rather spreading from ground, 10–150 cm long, dark green. Leaf-stalks with dark or dark-centred scales. Spore cases, **fringed with stalked glands**, close to main midrib of leaf pinnae.

> *Dryopteris dilatata*
> Broad Buckler Fern

spore cases
on pinnule

SECTION B

1   a Evergreen, woody climber or forming a carpet on woodland floor. **Sucker-like adhesive roots** present on stems. Leaves hairless, dark green above, paler below, 4–10 cm long, variable in shape, three to five palmately lobed or oval-elliptical, but not toothed. Flowers in umbels, petals yellowish-green. Fruits globular black.

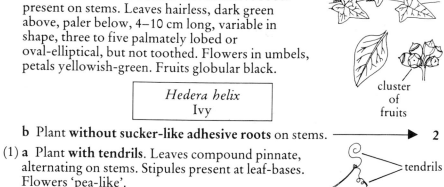

cluster of fruits

> *Hedera helix*
> Ivy

**   b** Plant **without sucker-like adhesive roots** on stems. ⟶ **2**

2 (1) a Plant **with tendrils**. Leaves compound pinnate, alternating on stems. Stipules present at leaf-bases. Flowers 'pea-like'.

tendrils

⟶ **3**

**   b** Plant **without tendrils**. ⟶ **5**

3 (2) a Leaves with **three or more pairs of leaflets.** ⟶ **4**

**   b** One pair of narrowly lanceolate leaflets, 7–15 cm long. Plant hairless, up to 3 m long. Stems broadly winged. Stipules 20 mm long, lanceolate, less than half the width of the stem. Flowers buff-yellow with upper parts tinged rose-pink, on three- to eight-flowered, stalked racemes. Open woods.

stipules

flower

> *Lathyrus sylvestris*
> Narrow-leaved Everlasting Pea

4 (3) a Leaves of three to nine pairs of **oval**, untoothed leaflets, 10–30 mm long. **Leaflet bases rather heart-shaped, tips blunt.** Stipules half-arrow-shaped. Plants 30–50 cm long. Flowers pink–purple in two- to six-flowered racemes.

> *Vicia sepium*
> Bush Vetch

**   b** Leaves of six to twelve (rarely fifteen) pairs of leaflets, 10–25 mm long, **oblong-lanceolate** to **linear-lanceolate**. Stipules entire, half-arrow-shaped. Plant more or less hairy, 60–200 cm long. Flowers purplish in 10- to 40-flowered racemes. Wood margins.

> *Vicia cracca*
> Tufted Vetch

9

5 (2) **a** Leaves **divided into distinct leaflets.** ————————▶ **6**
   b Leaves **not divided into distinct leaflets** but blades may be deeply cut into lobes. ————————————————————▶ **10**

6 (5) **a** Leaves **palmate or pinnate.** Prickles may be present. ———▶ **7**
   b Leaves three-foliate, leaflets oblong, toothed. **Stipules narrow, joined to leaf-stalk.** Plant 60–120 cm tall, erect, branched. Flowers 'pea-like', yellow, in racemes. In open woods, rarer in northern Britain.

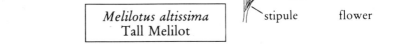

stipule   flower

> *Melilotus altissima*
> Tall Melilot

7 (6) **a** Leaves **pinnate.** ————————————————————▶ **8**
   b Leaves **palmate**, of three (rarely five) oval or oblong, toothed leaflets. Stipules present at leaf base. Stems woody, straight or curved. Straight or curved prickles and bristles present. Flowers 'rose-like' of five white or pink petals. Fruits green, red, then black 'blackberries'.

flower

fruit

> *Rubus fruticosus aggregate*
> Bramble

Rubus caesius (Dewberry) is similar but leaves of three, irregularly, coarsely **double-toothed** leaflets or leaflets shallowly lobed. Stems weak. Prickles scattered.

leaf margin

8 (7) **a** Leaves **alternate**, of two to three pairs of **toothed, oval leaflets. Stipules at base**. Stems green and weak, with arching prickles. Flowers 'rose-like' with five white petals. Fruits green, then red, oval 'rose-hips'.

> *Rosa arvensis*
> Field Rose

   b Leaves **without stipules at base of leaf-stalk, leaflets not toothed.**
————————————————————————————▶ **9**

9 (8) **a** Leaves opposite, usually with **five, narrow, oval, pointed leaflets**. Stems woody, fibrous and peeling. Flowers up to 20 mm across, of greenish-creamy sepals and many stamens, in terminal panicles in leaf-axils. Fruits develop long, white-plumed styles. Typically on calcareous soils.

> *Clematis vitalba*
> Traveller's Joy

b Leaves **alternate with a pair of leaflets** at the base of the main leaf-blade. Stems woody and downy, up to 2 m long. Flowers up to 10 mm across, in loose cymes. Corollas bell-shaped with five, arched-back, pointed, purple lobes and a central column of yellow stamens. Fruits red, oval berries.

> *Solanum dulcamara*
> Bittersweet

flower
(enlarged)

10 (5) **a** Leaves **arranged in whorls along four-angled stems.** ⟶ **11**
    b Leaves **not arranged in whorls along stems.** ⟶ **14**

11 (10) **a** Leaves **six to eight** in a whorl. ⟶ **12**
    b Leaves **four** in a whorl, oval-elliptical, **three-veined, very hairy**, yellow–green. Stems spreading then erect, 15–60 cm long. Flowers up to 2.5 mm across, four-lobed, pale yellow, in eight-flowered axillary cymes.

part of
stem

flower

> *Galium cruciata*
> Crosswort

12 (11) **a** Leaves linear-lanceolate or elliptical, 12–50 mm long, **one-veined** with **backwardly directed prickles** on margins. Stems up to 1 m long, very rough. Flowers four-lobed, whitish-green, up to 2 mm across, in two- to five-flowered axillary cymes.

part of
stem

> *Galium aparine*
> Cleavers

flower   fruit

    b Leaf **prickles forwardly directed.** ⟶ **13**

13 (12) **a** Leaves **lanceolate, one-veined**, 7–10 mm long. Stems smooth, sprawling. Flower corollas white.

> *Galium saxatile*
> Heath Bedstraw

b Leaves **firm, in distant whorls, lanceolate** or **elliptical. Stems typically erect**, hairy beneath nodes, otherwise hairless.

> See *Galium odoratum*, page 47

14 (10) **a** Leaves **alternate, entire** or **lobed**, up to 8 cm long.

> See *Solanum dulcamara*, page 11

b Plant **without the combined features** of *Solanum dulcamara*. ⟶ **15**

15 (14) **a** Margins of leaves **more or less entire**, not toothed, not deeply cut into lobes. ⟶ **16**
b Leaves **opposite**, long-stalked, **palmately lobed, toothed**, up to 15 cm long. Plant square-stemmed, up to 3 m long. Male flowers in branched yellow-green clusters. Female flowers are catkins resembling small pine cones.

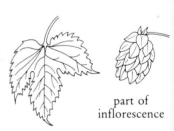

part of
inflorescence

> *Humulus lupulus*
> Hop

16 (15) **a** Leaves **alternate**. ⟶ **17**
b Leaves **opposite**, 30–70 mm long, **oval to elliptical**, grey-green. Lower leaves short-stalked. Plant woody, up to 6 m tall. Flowers in whorled, terminal heads, corollas trumpet-shaped, yellow to purplish-pink on outside, very fragrant. Fruits red, globular berries.

> *Lonicera periclymenum*
> Honeysuckle

17 **(16) a** Leaves **oval-triangular** or **heart-shaped.** ⟶ **18**
     b Leaves rather oblong, **arrow-shaped**, up
to 15 cm long. Plant up to 3 m tall. Flowers
trumpet-shaped, solitary. Corollas white or
pink, up to 40 mm wide.

flower

> *Calystegia sepium*
> Hedge Bindweed

18 **(17) a** Leaves 2–6 cm long, **oval-triangular** or
heart-shaped with shortened bases, longer
than leaf-stalks. Stems angular, up to 1 m or
more in length. Flowers in long racemes
with three greenish-grey outer sepals with
white keels or narrow wings in fruit.

> *Fallopia (Polygonum) convolvulus*
> Black Bindweed

     b Leaves 3–10 cm long, **heart-shaped**, pointed,
hairless, long-stalked, **very glossy dark green.**
Stems slender, angled, climbing and twisting
clockwise, 2–4 m long. Flowers bell-shaped, six-
lobed, yellow-green, in racemes in leaf-axils.

> *Tamus communis*
> Black Bryony

SECTION C

1 Woody shrubs or trees with **prickles, thorns or spines.**
                           ⟶ Section C1, page 13
2 Trees with **needle-like leaves.** ⟶ Section C2, page 15
3 Shrubs or trees with **compound palmate or pinnate leaves.**
                           ⟶ Section C3, page 16
4 Shrubs with **simple or palmately lobed, opposite pairs of leaves.**
                           ⟶ Section C4, page 17
5 Shrubs or trees with **simple, alternate leaves.** ⟶ Section C5, page 19

Section C1
With prickles, thorns or spines
1   **a** Stems with 'rose-like' prickles. ⟶ **2**
      b Stems with **woody thorns or spines.** ⟶ **3**

2 (1) **a** Leaves **palmate** of three irregularly, coarsely double-toothed leaflets or leaflets shallowly lobed.

> See *Rubus caesius*, on page 10

b Leaves **pinnate** of two to three pairs of toothed leaflets. Stems strong and arching, up to 3 m tall, with broad-based, strongly hooked prickles. Flowers 'rose-like', 30–50 mm across, of five white or pink petals. Fruits red or scarlet 'rose-hips'.

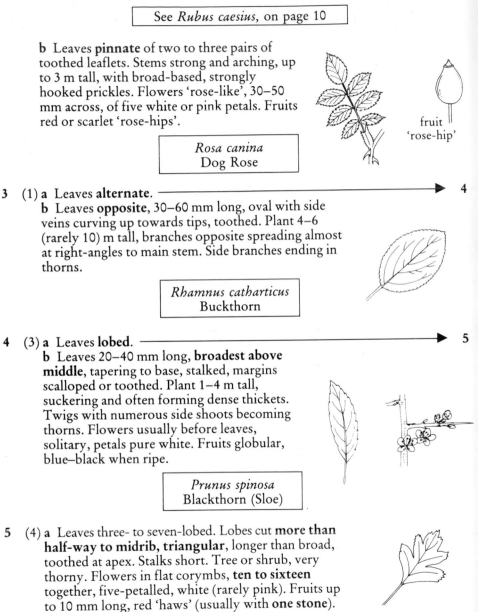

fruit 'rose-hip'

> *Rosa canina*
> Dog Rose

3 (1) **a** Leaves **alternate**. ────────────▶ **4**

b Leaves **opposite**, 30–60 mm long, oval with side veins curving up towards tips, toothed. Plant 4–6 (rarely 10) m tall, branches opposite spreading almost at right-angles to main stem. Side branches ending in thorns.

> *Rhamnus catharticus*
> Buckthorn

4 (3) **a** Leaves **lobed**. ────────────▶ **5**

b Leaves 20–40 mm long, **broadest above middle**, tapering to base, stalked, margins scalloped or toothed. Plant 1–4 m tall, suckering and often forming dense thickets. Twigs with numerous side shoots becoming thorns. Flowers usually before leaves, solitary, petals pure white. Fruits globular, blue–black when ripe.

> *Prunus spinosa*
> Blackthorn (Sloe)

5 (4) **a** Leaves three- to seven-lobed. Lobes cut **more than half-way to midrib, triangular,** longer than broad, toothed at apex. Stalks short. Tree or shrub, very thorny. Flowers in flat corymbs, **ten to sixteen** together, five-petalled, white (rarely pink). Fruits up to 10 mm long, red 'haws' (usually with **one stone**).

> *Crataegus monogyna*
> Hawthorn

b Leaves three- to seven-lobed. Lobes cut to **less than half-way to midrib, rounded,** toothed nearly all round. Flowers in corymbs, usually **ten or fewer** together, five-petalled, white. Fruits up to 10 mm long, deep red 'haws' (usually with **two stones**).

> *Crataegus laevigata*
> Midland Hawthorn

Section C2
Needle-like leaves

1 **a** Leaves **arranged in bunches of two.** ⟶ 2
   b Leaves **not arranged in bunches of two.** ⟶ 3

2 (1) **a** Leaves 30–80 mm long, stiff, hairless. Trees 30 m or more tall. Bark furrowed and scaly, bright reddish-brown in upper parts. Male and female flowers in separate cones.

> *Pinus sylvestris*
> Scots Pine

b Leaves hard-pointed, four-angled, with inconspicuous lines on each side. Tree 36 m or more tall (often cut when small). Bark reddish-brown, flaking into thin scales or plates.

> *Picea abies*
> Norway Spruce (Christmas Tree)

3 (1) **a** Leaves arranged **in bunches of more than five.** ⟶ 4
   b Leaves arranged **in whorls of three or in two rows.** ⟶ 5

4 (3) **a** Leaves up to 35 mm long, **bright emerald green** (fading to a pale straw colour before falling), either scattered around young shoots or in rosettes of twenty to thirty on older twigs. Tree 30–42 m tall.

> *Larix decidua*
> Common Larch

b Leaves with **dark green upper surface and two bands of grey.** Bark dark reddish-brown, cracked or broken into scales. Scales on cones are back-turned and give the cone the appearance of a rose. Tree 25–30 m tall.

> *Larix kaempferi*
> Japanese Larch

5 (3) **a** Leaves 5–10 mm long, tapering to sharp points, arranged **in whorls of three on twigs**. Shrub up to 10 m tall. Bark reddish-brown. Cones small, male up to 8 mm, female up to 2 mm.

> *Juniperus communis*
> Juniper

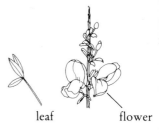

b Leaves **in two rows**, 10–30 mm long, strap-shaped, dark green above, paler below, midrib prominent on both sides. Tree up to 20 m tall, trunk massive. Seeds surrounded by red, fleshy cups.

> *Taxus baccata*
> Yew

Section C3
Compound palmate or pinnate leaves

1 **a** Leaves **of one to three leaflets**, stalked, on younger stems. Twigs usually **with five raised angles**. Flowers 'snapdragon-like', golden-yellow. Plant found in scrub on dry acid soils.

> *Cytisus scoparius*
> Broom

leaf   flower

b Leaves **compound palmate, opposite, of five to seven leaflets**, 8–20 cm long, leaflets broadest above middle and tapering to base, pointed at tip, toothed. Tree up to 25 m tall with dark grey–brown bark. Twigs stout, curved up at ends. Buds large, sticky, oval, red–brown, hairless. Inflorescences large, erect terminal panicles. Flowers up to 20 mm across, of four white (spotted pink) petals. Seeds 'conkers' in globular green (later brown) prickly fruits.

> *Aesculus hippocastanum*
> Horse Chestnut

c Leaves **compound pinnate**. **2**

16

2 **(1) a** Leaves **alternate**, usually with **6–7 leaflets** 30–60 mm long (terminal no longer than side leaflets), **strongly toothed**, dark above, whitish–grey and downy below. Tree usually up to 15 m tall. Bark smooth, shining (grey–brown). Twigs greyish–brown with oval, dark brown buds. Flowers five-petalled, creamy-white in dense corymbs resembling umbels. Fruits globular, scarlet. Dry woods on acid soils, especially in northern Britain.

> *Sorbus aucaparia*
> Rowan

b Leaves **opposite**. ──────────────────────▶ **3**

3 **(2) a** Leaves of **seven to thirteen** shallow-toothed, oval-pointed leaflets up to 70 mm long. Tree up to 30 m tall. Bark **pale grey**. Twigs grey, buds black. Flowers in panicles (appearing before leaves), purplish. Fruits oblong, winged at tips (like a propeller blade). Base-rich soils.

fruits

> *Fraxinus excelsior*
> Ash

b Usually **one to two pairs of side leaflets**, leaflets 30–90 mm long, oval or elliptical, toothed, more or less hairless, with distinct, unpleasant smell. Shrub or small tree up to 10 m tall, may have erect suckers arising from base. Bark **deeply furrowed, corky**. Inflorescences umbel-like, flat-topped, much-branched. Corollas of flowers five-lobed, creamy. Fruits purple–black berries in clusters.

> *Sambucus nigra*
> Elder

Section C4
Simple or palmately lobed, opposite pairs of leaves

1 **a** Leaves **toothed** (may be very finely so), may be palmately lobed. ▶ **2**
b Leaves **untoothed**, 40–80 mm long, oval, pointed, wavy-edged, rounded at base, slightly downy on both sides. **Three to five main veins on each side curve round towards leaf apex.** Shrub up to 4 m tall. Twigs purplish-red. Flowers four-petalled, creamy-white, in flat-topped stalked umbels. Fruits black 'berries'. On calcareous soils.

> *Thelycrania sanguinea*
> Dogwood

2 (1) **a** Leaves **simple.** ➔ 3

    b Leaves **palmately lobed.** ➔ 4

3 (2) **a** Leaves 5–10 cm long, **wrinkled**, pointed, finely toothed, densely grey downy below. Plant 2–6 m tall, twigs downy, buds **without enclosing scales.** Inflorescences flat, umbel-like heads, 6–10 cm across. Flowers **three-petalled**, creamy, all alike.

> | *Viburnum lantana* |
> |---|
> | Wayfaring Tree |

    b Leaves 3–13 cm long (orange to red in autumn), **oval-lanceolate**, finely toothed. Shrub or small tree 2–6 m tall. Twigs green, hairless, **four-angled.** Flowers with **four greenish-white petals** and four stamens alternating with petals, in stalked, forked cymes. Fruits four-lobed, deep pink.

flower

> | *Euonymus europaeus* |
> |---|
> | Spindle Tree |

4 (2) **a** Leaves 50–80 mm long with **three to five, irregularly, sharply toothed lobes**, dark green and more or less downy below. Shrub 2–4 m tall. Twigs greyish, slightly angled, buds scaly. Flowers white, five-petalled, inner flowers up to 6 mm across, outer flowers up to 20 mm across, in flat, umbel-like heads. Fruits globular, shiny red.

> | *Viburnum opulus* |
> |---|
> | Guelder Rose |

    b Leaves 7–16 cm long, **five-lobed (lobes to about half-way to midrib)**, hairless, blunt-toothed, dark green above, paler below. Tree up to 30 m tall. Bark smooth (flaking when old). Buds oval, scales green with black edges. Inflorescences narrow, drooping panicles of yellowish-green flowers. Fruits winged, wings spreading at acute angles.

> | *Acer pseudoplatanus* |
> |---|
> | Sycamore |

Acer campestre (Field Maple) is similar to the above but smaller, rarely more than 15 m tall. The leaves are **three- to five-lobed and cut to about half-way** with heart-shaped base, hairs on veins below. Twigs often develop corky wings.

Section C5
Simple, alternate leaves

1 a Leaves oval, 3–10 cm long, margins bearing **large spine-pointed teeth**, evergreen, dark glossy green, leathery. Leaves at the top often spineless. Shrub or small tree 5–15 m tall. Bark grey, smooth, eventually fissured. Twigs green. Flowers white, four-petalled. Fruits red berries.

> *Ilex aquifolium*
> Holly

b Leaves **without spine-pointed teeth.** ──────▶ 2

2 (1) a Leaf margins **entire.** ──────▶ 3
b Leaf margins **lobed, wavy-edged or toothed.** ──────▶ 4

3 (2) a Leaves broadest above middle, 20–70 mm, shiny green (yellow–red in autumn), with brownish hairs when young. Lateral veins large (about seven pairs). Shrub or small tree 4–5 m tall. Flowers small of five greenish-white petals, in clusters in axils of upper leaves. Fruits green then red, finally violet–black berries.

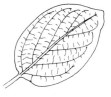

> *Frangula alnus*
> Alder Buckthorn

b Leaves **towards upper part of plant,** lanceolate, 5–12 cm long, dark green, evergreen. Shrub up to 1 m tall with little-branched stems. Stems erect. Flowers green in short, axillary, five- to ten-flowered, racemes. Mainly on calcareous soils, especially in southern England.

> *Daphne laureola*
> Spurge Laurel

4 (2) a Leaves **lobed.** ──────▶ 5
b Leaves **wavy-edged** or **toothed.** ──────▶ 6

5 (4) **a** Leaves **oblong**, 5–12 cm long, dull green with rounded side lobes. **Basal lobes (auricles) rounded and exceeding the short leaf-stalk.** Tree up to 30 m or more tall. Bark brownish-grey, fissured. Twigs greyish-brown. Male catkins 2–4 cm long, stamens six to eight per flower. Female spikes one- to five-flowered. Fruits acorns, two to three together on long stalks.

> *Quercus robur*
> Pedunculate Oak

b Leaves less lobed, **wedge-** or **heart-shaped, without auricles. Star-shaped hairs in axils** of side veins beneath. Acorns on very short stalks. On poorer soils.

> *Quercus petraea*
> Sessile Oak

6 (4) **a** Leaves **broadly** or **narrowly lanceolate.** 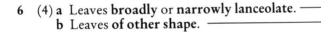 → 7
   b Leaves **of other shape.** → 9

7 (6) **a** Leaves 10–25 cm long, **broadly lanceolate,** pointed, with **large pointed teeth**. Tree up to 30 m tall. Bark dark, brownish-grey, vertically fissured, often spiralling round trunk. Twigs olive–brown. Catkins 12–20 cm long with yellowish–white anthers. Fruits brown in green, large-spined cups.

> *Castanea sativa*
> Sweet Chestnut

b Leaves **narrowly lanceolate.** Plant **without the above features.** → 8

8 (7) **a** Leaves 5–10 cm long, **finely toothed,** covered **with white silky hairs.** Tree 10–25 m tall, branches **ascending at 30–50°.** Bark rugged, greyish, fissured. Twigs silky when young, later hairless and olive. Male catkins up to 5 cm long, flowers with two yellowish anthers. In wet woods on richer soil.

> *Salix alba*
> White Willow

b Leaves 6–15 cm long, **coarsely toothed, without silky hairs**. Stalk with two glands at top. Tree 10–25 m tall, branches **spreading widely at 60–90°**. Bark greyish, deeply fissured. Twigs olive, very fragile. Male catkins up to 6 cm long, drooping, flowers with two yellowish anthers. In wet woods, more tolerant of poor soils.

> *Salix fragilis*
> Crack Willow

9 (6) **a** Leaf-blade extending **farther down stalk on one side than on the other**. Leaves 4–9 cm long, **rounded-oval, roughly hairy**. Leaf-stalks very short. Trees up to 30 m tall, few large branches in lower part, suckers often abundant. Catkins tassel-like, purplish. Trees greatly reduced by Dutch Elm Disease.

> *Ulmus procera*
> English Elm

Ulmus glabra (Wych Elm) occurs in hilly, rocky woods. Leaves are 8–12 cm long, **oval-diamond** to **elliptical, very rough**. Leaf-stalks very short.

b Leaf-blade **not obviously extending farther down stalk on one side than on the other**. ⟶ 10

10 (9) **a** Leaves **triangular**, 25–50 mm long, **double-toothed**, hairless. Tree up to 25 m tall. Bark silver–white, papery. Twigs hairless, warty. Male catkins drooping, 30–60 mm long. Female catkins erect, shorter, scales three-lobed. Fruits with rounded, papery wings.

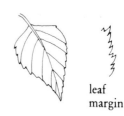

leaf
margin

> *Betula pendula*
> Silver Birch

Betula pubescens (Downy Birch) has **oval-triangular** leaves which are **single-toothed** and downy below. Bark brownish, peeling and papery.

b Leaves **of other shape**. ⟶ 11

.**11** (10) **a** Leaves **oval or oval-elliptical**. ⟶ 12
    b Leaves **of other shape**. ⟶ 16

12 (11) **a** Leaves oval, margins **typically wavy-edged**. ⟶ 13
    b Leaves oval or oval-elliptical, **margins toothed**. ⟶ 14

13 (12) **a** Leaves 4–9 cm long, stalked. **Veins prominent at edges**, edges and veins beneath **silky-hairy**. Tree 30 m or more tall. Bark **smooth, grey**. Twigs red–brown, buds **cigar-shaped**. Male catkins tassel-like on long stalks. Female flowers in pairs on long stalks, surrounded by a scaly cup. On chalk, limestone, sands and light loams.

Fagus sylvatica
Beech

b Leaves 5–10 cm long, pointed tips, hairless above, **grey downy below**. Shrub 3–10 m tall. Bark **coarsely fissured**. Twigs downy when young, rather stout, buds **three-angled**. Catkins dense, oblong-oval, appearing before leaves.

Salix caprea
Goat Willow

14 (12) **a** Leaves oval-elliptical, double-toothed, 6–15 cm long, pointed, hairless above, downy below. **Two red glands on stalk just below leaf-blade**. Trees up to 25 m tall, suckering freely. Bark smooth, reddish–brown, peeling. Flowers cup-shaped, of five white petals. Fruits bright or dark red 'cherries'.

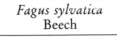

gland
on
leaf-stalk

Prunus avium
Wild Cherry

b Leaves oval, double-toothed. ⟶ **15**

15 (14) **a** Leaves 5–12 cm long, variable, sides mostly gradually curved, **white woolly below**. Trees up to 25 m tall. Bark dark grey, **shallowly fissured**. Twigs chestnut–brown to dark grey. Buds oval, greenish. Petals of flowers white, oval, in compound corymbs. On chalk and limestone.

Sorbus aria aggregate
Whitebeam

b Leaves 3–10 cm long, pointed, **with pleated side veins**, hairless above, **hairy on veins below**. Tree up to 30 m tall (usually much less). Bark smooth grey, **trunk angled**. Twigs downy. Male catkins up to 5 cm long with oval, greenish bracts. Female catkins up to 20 mm long (10 cm in fruit). Fruits oval, ribbed nuts. In woods, especially on loamy and sandy clays.

> *Carpinus betulus*
> Hornbeam

16 (11) **a** Leaves **broadest above middle.** ⟶ **17**
　　　　b Leaves **not broadest above middle**, rounded in outline. ⟶ **19**

17 (16) **a** Leaves 20–30 mm long, rounded, pointed, very wrinkled, grey woolly below, with **conspicuous, kidney-shaped stipules at base**. Shrub 1–3 m tall with numerous spreading branches. Twigs brown, usually angular. Catkins appearing before leaves, males 1–2 cm long, oval. In damp woods.

> *Salix aurita*
> Eared Willow

b Leaves may have **stipules** (but they may not persist) or **no stipules**.
⟶ **18**

18 (17) **a** Leaves 20–60 mm long, **tapering to stalk**, dark grey–green above, downy below. Shrub 2–10 m tall, branches long, straight and nearly erect. Twigs stout, brown. Buds **oval**. Catkins dense, oval, appearing before leaves.

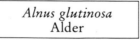

> *Salix cinerea aggregate*
> Grey Willow

b Leaves 30–90 mm long, rounded, **appearing cut across at tip**, often double-toothed, dark green, hairless except on veins below. Tree up to 20 m tall. Bark dark, fissured. Buds **blunt**, purplish with large outer scale. Male catkins 20–60 mm long, pendulous, yellow with dark tips to bracts. In wet places.

> *Alnus glutinosa*
> Alder

19 (16) **a** Leaves **double-toothed**, 5–10 cm long, **tip drawn out into a point**, base heart-shaped. Leaf-stalk hairy. A many-stemmed shrub up to 8 m tall. Bark coppery-brown, smooth, tendency to peel. Twigs very hairy with reddish, gland-tipped hairs. Catkins appear before leaves. Male catkins 20–80 mm long, pendulous, anthers yellow. Female catkins up to 5 mm long, resembling buds, styles reddish.

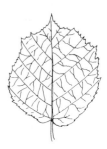

> *Corylus avellana*
> Hazel

b Leaves **not double-toothed.** ⟶ **20**

20 (19) **a** Leaves 6–10 cm long, **long-stalked,** extended into a **distinct point, tufts of white hairs** on the veins below. Tree up to 25 m tall. Trunk with large irregular bumps (bosses) on surface. Inflorescences of four to ten small flowers on common stalk with an oblong leaf-like bract. Fruits nut-like, globular.

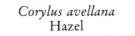

portion of under-side of leaf showing tufts of hairs

> *Tilia europaea* (a hybrid)
> Common Lime

Tilia cordata (Small-leaved Lime) has **shorter** leaves with **abrupt points**. Veins below have **tufts of rusty-coloured hairs**.

b Leaves 20–60 mm long **with four to six blunt, single triangular teeth** on each side, **white or woolly below**. Leaf-stalks **short**. Trees up to 35 m tall. Bark smooth, grey with rhomboidal lenticels. Catkins up to 10 cm long, stout, stigmas yellow. In damp and wet woods in lowland areas.

> *Populus canescens*
> Grey Poplar

Populus tremula (Aspen) has leaves **with eight to ten large, rounded teeth** on each side. Leaf-stalks **long and very flexible**. More typical of northern and western Britain.

SECTION D

1   **a** Leaves **mainly basal**, grass-like, limp, with **long hairs scattered over the surface**, sheathing bases very short. Inflorescences in spreading, forking clusters.

| Woodrushes | ➤ 2 |

   b Plant **without the combined features of the above.** ➤ 3

2   (1) **a** Basal leaves 10–30 cm long, **up to 12 mm or more wide**, glossy, tapering to a fine point. Plant robust, forming bright green **mats** or **tussocks**. Stems up to 80 cm tall with numerous runners. Inflorescence a widely spreading, forking cluster with many small heads of two to five brown flowers.

| *Luzula sylvatica*
 Great Woodrush |

   b Leaves **up to 6 mm wide**. Stems up to 40 cm tall, wiry. Inflorescence a cluster of three to ten stalked heads each with eight to eighteen red–brown flowers, anthers about **as long as** the filaments.

| *Luzula multiflora*
 Many-headed Woodrush |

   c Leaves **up to 10 mm wide**. Stems up to 30 cm tall. Inflorescence with single flowers on branches, anthers **longer than** filaments.

| *Luzula pilosa*
 Hairy Woodrush |

3   (1) **a** Leaves **narrow-linear** in **two** vertical ranks. Stems **hollow** and **round**.

| Grasses | ➤ 4 |

   b Leaves **long-linear, hairless**, in **three** vertical ranks. Stems **solid**, more or less **triangular** in cross section.

| Sedges | ➤ 12 |

25

4 (3) **a** Leaves **narrow and bristle-like**, 0.3–0.8 mm wide, rough. Plant loosely to densely tufted with a slender, erect or bent stem, 25–90 cm tall. Inflorescence spreading, 5–10 cm long, usually purplish. Spikelets 4–6 mm long, purplish, on long stalks. In open woodlands.

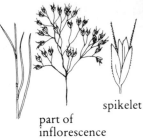

spikelet

part of inflorescence

> *Deschampsia flexuosa*
> Wavy Hair-grass

Other smaller, bristle-leaved grasses such as *Festuca rubra* (Red Fescue) may be found especially in open birch woods.

b Leaves **with a distinct blade, not bristle-like.** ⟶ **5**

5 (4) **a** Leaf-sheath **with a short bristle** on the side opposite the leaf-blade. Leaves flat, narrowed to a fine point, 5–20 cm long, 3–7 mm wide, bright green, rather thin, shortly hairy above. Plant forming loose, leafy patches, 20–60 cm tall, with slender, white, creeping rhizomes. Inflorescences very loose, sparingly branched, erect or nodding, 6–22 cm long, 1–12 cm wide. Spikelets elliptical or oblong, 4–7 mm long, without awns.

bristle

spikelet

inflorescence

> *Melica uniflora*
> Wood Melick

b Leaf-sheaths **without a bristle.** ⟶ **6**

6 (5) **a** Stems **conspicuously hairy at the nodes**. Leaves greyish–green, pointed, flat, 4–20 cm long, 3–12 mm wide. Plant 20–100 cm tall, forming compact tufts or loose mats, with tough, creeping rhizomes. Inflorescences narrowly oval or oblong spikes, compact to somewhat loose, 4–12 cm long. Spikelets elliptical or oblong, flattened, 4–6 mm long, with awns. Open woods, often carpeting the ground.

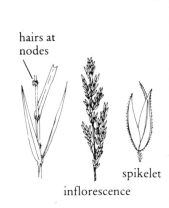

hairs at nodes

spikelet

inflorescence

> *Holcus mollis*
> Creeping Soft-grass

Holcus lanatus (Yorkshire Fog) is similar but sheaths and leaves **softly hairy throughout**.

b Stems **not conspicuously hairy at nodes.** ⟶ **7**

7 (6) **a** Leaf-sheaths **with a conspicuous beard of hairs** around the sheath junction, or sheaths and blades **hairy.** ⟶ **8**
  b Leaf-sheaths **not hairy, no beard of hairs** at the junction. ⟶ **9**

8 (7) **a Beard of hairs** around the sheath junction with the leaf-blade, ligule 1–5 mm long. Leaves green, finely pointed, 1–12 cm long, 1.5–5.0 mm wide (but variable, may be longer and wider in wet places). Plant 10–100 cm tall, tufted, when bruised, smelling of new-mown hay. Inflorescences very dense to somewhat loose, oval to narrowly oblong, 1–12 cm long, 6–15 mm wide, green or purplish. Spikelets lanceolate, compressed, 6–10 mm long, **awns small.**

ligule
beard of hairs
spikelet
inflorescence

> *Anthoxanthum odoratum*
> Sweet Vernal Grass

b Leaf-sheaths and blades **hairy,** ligules blunt, 1–6 mm long. Leaves green, blades narrowed to sheath, finely pointed, flat, up to 35 cm long, 4–12 mm wide. Plant 30–90 cm tall, compactly tufted. Inflorescences rather spike-like, loose, erect or nodding, 6–20 cm long, bearing four to twelve spikelets. Spikelets 20–40 mm long, cylindrical, lanceolate or narrowly oblong, **alternating in two rows on opposite sides of the axis, long awned.**

ligule
inflorescence   spikelet

> *Brachypodium sylvaticum*
> Wood False Brome

9 (7) **a** Grass **without rhizomes or stolons.** ⟶ **10**
  b Grass **with slender creeping stolons, rooting at nodes.** Leaves bright green to grey–green, up to 3 mm wide, finely pointed, 2–15 cm long, flat or rolled, soft. Plant loosely tufted, 15–75 cm tall. Inflorescences rather loose and open, 3–16 cm long, up to 70 mm wide, erect or nodding, purple, reddish or green, branches 'hair-like'. Spikelets lanceolate to narrowly oblong, 1.5–3.0 mm long with a single, bent awn (may be absent). Damp and wet places in woods.

> *Agrostis canina*
> Velvet Bent

part of inflorescence   spikelet

Agrostis tenuis (Common Bent) is similar but **with short stolons.**

10   (9) **a** Apex of leaf-sheaths **with prominent spreading, narrow collars (auricles).** Sheath and leaf-blade junction typically dark purple to red. Leaves bright green, long, tapering to a fine tip, flat, up to 60 cm long, 6–18 mm wide. Plant 45–190 cm tall, loosely tufted. Inflorescences nodding, loose, oval to lanceolate in outline, 10–50 cm long, green. Spikelets lanceolate to narrowly oblong, 8–20 mm long, long awned.

Festuca gigantea
Giant Fescue

   b Apex of leaf-sheath **without auricles.** ──────▶ **11**

11   (10) **a** Grass **forming large tussocks,** ligules up to 15 mm long. Leaves usually channelled, green, sharply pointed or blunt, 10–60 cm long, 2–5 mm wide, margins coarse. Plant 20–200 cm tall. Inflorescences erect or nodding, **oval** to **oblong in outline,** 10–50 cm long, up to 20 cm wide, variously coloured silvery, golden, purple, green. Spikelets **loosely scattered or clustered,** lanceolate to narrowly oblong, 4–6 mm long, awned.

Deschampsia caespitosa
Tufted Hair-grass

   b Grass **not forming tussocks,** ligules short, 0.5–1.5 mm long. Leaves bright green, blades finely pointed, flat, 10–30 cm long, 4–13 mm wide, rather thin, rough. Plant 30–110 cm tall, loosely tufted. Inflorescences **spike-like,** curved or nodding, slender, 5–20 cm long, green or tinged purple. Spikelets **alternating in two rows on opposite sides of axis,** long awned.

Agropyron caninum
Bearded Couch

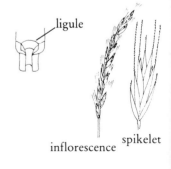

12 (3) **a** Sedge **forming substantial tussocks or clumps.** Stems **may be over** 100 cm tall. ━━━━━━━━━━━━━━━━━━━━━━━━━━━▶ **13**

b Sedge **tufted** (may be **creeping**). Stems **never reaching** 100 cm tall.
━━━━━━━━━━━━━━━━━━━━━━━━━━━━━━━▶ **14**

13 (12) **a** Leaves **dark green, rough with upward directed prickles,** up to 120 cm long, 5–7 mm wide. Stems up to 150 cm tall, rough, in **substantial tussocks,** of three flat sides and dark brown scales at bases. Inflorescences 5–15 cm long, oblong, brown **spike.** Male and female flower-heads close together. Bract bristle-like, shorter than whole spike.

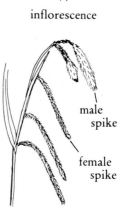

> *Carex paniculata*
> Tussock Sedge

inflorescence

b Leaves broad, 15–20 mm wide, **rough-edged, green** or **yellow–green above,** bluish below, shorter than stems. Stems up to 180 cm tall, bluntly three-angled, growing in **clumps.** Inflorescences **drooping,** one to two male spikes above four to five, narrowly cylindrical female spikes, 7–16 cm long. Lower bracts shorter than inflorescence.

> *Carex pendula*
> Drooping Sedge

male
spike

female
spike

inflorescence

14 (12) **a** Leaves **very narrow, 2 mm (or less)** wide, pale green, channelled, roughish at margins. Plant tufted, stems up to 60 cm tall, weak three-sided. Inflorescences 8–20 cm long, lower flower-spikes well-spaced-out, pale green, males at top. Leaf-like bracts long.

> *Carex remota*
> Remote Sedge

inflorescence

b Leaves typically **more than 2mm** wide. Sedge **without the combined features of the above.** ━━━━━━━━━━━━━━━━▶ **15**

15 **(14) a** Leaves grooved, 2–5 mm wide, **with long white hairs on leaf-sheaths**. Stems hairless, shiny, three-angled. Plant tufted, creeping, up to 70 cm tall. Inflorescences with two to three long, fattish female spikes spaced well down the stem. Male spikes two to three, rather slender and inconspicuous with bristle-like bracts.

> *Carex hirta*
> Hairy Sedge

inflorescence

b Leaves typically **not hairy**. Sedge **without the combined features of the above**. ——————————————————————▶ **16**

16 **(15) a** Leaves short, 3–6 mm wide, **limp, shiny green** or **yellow–green**. Plant tufted. Stems up to 60 cm tall, smooth, three-sided. Inflorescences nodding, one male spike above and three to five long-stalked, long and thin female spikes. Lower bracts erect, sometimes overtopping inflorescences.

> *Carex sylvatica*
> Wood Sedge

inflorescence

b Leaves stiff, **greyish, bluish beneath**, 2–6 mm wide, **shorter than stems**. Plant loosely tufted, far creeping. Stems smooth, bluntly three-sided. Inflorescences with spikes close at top, one to three male spikes above one to five, long, stalked, fattish, drooping female spikes. Lowest bract about as long as inflorescences.

> *Carex flacca*
> Glaucous Sedge

inflorescence

SECTION E

1   **a**   Plants with leaves arising **at** ground level.
Stem leaves may or may not be present.
→   Section E1, page 31

      b   Plants with leaves arising only from stems **above** ground level. →   **2**

2   (1) **a**   Overwintering first-year plant bearing **terminal rosettes** of strap-shaped, downy, dark green leaves, 30–80 mm long, short-stalked. Flower shoots arise from rosettes in second year, leaves similar, **alternate**. Flowers green–yellow in cup-shaped structure of crescent-shaped parts, in five- to ten-rayed umbels.

> *Euphorbia amygdaloides*
> Wood Spurge

      b   Plants with **all leaves in opposite pairs**, or **only lower leaves in opposite pairs** and **upper leaves** either **alternate** or **in apparent whorls of three**.
→   Section E2, page 39

3   **a**   Plants with **all leaves alternate**.
→   Section E3, page 47

Section E1

1   **a**   Leaves with **parallel** veins. ————→ **2**
      b   Leaves with **netted** veins. ————→ **7**

2   (1) **a**   Leaves **all** basal. ————→ **3**
      b   Leaves basal and **also present on flower-stems**. Flowers 'orchid-like'.
————→ **5**

3   (2) **a**   Leaves long-stalked, oval-elliptical, untoothed, 8–20 cm long, 30–50 mm wide, **arising in pairs from creeping rhizome**. Plant hairless, 10–20 cm tall. Inflorescence stalks leafless, flowers white, six-toothed, very fragrant, in one-sided racemes.

> *Convallaria majalis*
> Lily of the Valley

      b   Leaves **not arising from rhizomes**. ————→ **4**

4 (3) **a** Leaves **linear**, 20–40 cm long, glossy green, tips hooded, 10–20 mm wide, unspotted. Flower-stem hairless, 20–50 cm tall. Inflorescence a one-sided raceme of drooping, cylindrical, star-shaped flowers. Flowers sky-blue (sometimes white), **six-lobed**, anthers cream.

> *Hyacinthoides (Endymion) non-scriptus*
> Bluebell

b Leaves two to three, **elliptical-oval**, 10–25 cm long, bright green, pointed, on long stalks, **twisted through 180°**. Stems 10–45 cm tall, weakly three-angled. Inflorescence six- to twenty-flowered, flat-topped. Flowers white, **five-lobed**. Garlic- or onion-smelling when bruised.

inflorescence

> *Allium ursinum*
> Ramsons

5 (2) **a** Leaves **usually with purple blotches or spots**. ⟶ **6**
b Leaves **unspotted**. Basal leaves more or less opposite, elliptical-oblong, blunt, 10–15 cm long, rising at an angle from the base. Stem leaves, one to five, small, rather linear. Plant 30–40 cm tall. Flower spike 5–20 cm long, loose. Flowers 18–25 mm wide, greenish tinged creamy. Bracts long and leafy.

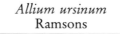

> *Platanthera chlorantha*
> Greater Butterfly Orchid

6 (5) **a** Leaves **alternating from basal rosette up stem**, 5–20 cm long, broadly to narrowly oblong, 2–3 cm wide, shiny, dark green, usually with dark purple blotches (arranged lengthwise). Stem leaves **sheathing stem**. Stem solid, rather stout, 20–40 cm tall. Inflorescence a loose raceme, 5–15 cm long. Flowers **bright purple–crimson**.

> *Orchis mascula*
> Early Purple Orchid

b Rosette leaves broad elliptical, blunt, grey–green, **with transversely elongated purple spots**. Stem leaves **not sheathing stem**, narrowly lanceolate, pointed, merging in size into leafy flower bracts. Stem solid, 15–40 cm tall. Flower spike conical, up to 10 cm long. Flowers **pale pink with purple streaks and spots**.

> *Dactylorhiza fuchsii*
> Common Spotted Orchid

7 (1) **a** Basal leaves **arrow-shaped**, 7–20 cm long, thin, very shiny, bright green, often purple spotted, more or less wrinkled, on long stalks. Plant 30–50 cm tall. A large, cup-shaped, leaf-like bract (spathe) surrounds inflorescence. Spathe erect, pointed, pale yellow–green, purple-edged, sometimes spotted. Inflorescence (spadix) 7–12 cm long, upper part cylindrical, club-shaped, usually pale purple. Fruits red berries in dense spike.

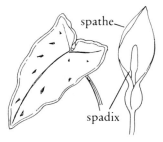

> *Arum maculatum*
> Lords and Ladies

b Basal leaves **not arrow-shaped**. Plant **without the combined features of the above.** ⟶ **8**

8 (7) **a** Leaves in basal rosette or arising from rootstock, **no stem leaves**. ⟶ **9**

b Leaves **present on stems.** ⟶ **11**

9 (8) **a** Leaves arising from rootstock, oval or rounded with **heart-shaped bases**, 20–60 mm long with **dense spreading hairs**, margins wavy to toothed, long-stalked. Stipules lanceolate. Flowers **'violet-like'**, **pale violet** to **white**, spur often upwardly curved or hooked.

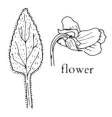

flower

> *Viola hirta*
> Hairy Violet

Viola odorata (Sweet Violet) has a thick rhizome and long, rooting stolons. Flowers scented, **deep violet** in colour.

b Leaves **in basal rosette**. Flowers not 'violet-like'. ⟶ **10**

10 (9) **a** Leaves 8–15 cm long, **spoon-shaped**, more or less stalked, **very wrinkled**, irregularly toothed, broadest above middle, downy below. Stems **woolly**. Flowers upright, wheel or saucer-shaped, 3–4 cm across, pale yellow, five shallow-notched lobes.

flower

> *Primula vulgaris*
> Primrose

b Leaves **heart-shaped**, hairless, wavy-edged, on long stalks. Stems **branched**, 5–25 cm tall. Flowers 'buttercup-like'. Petals eight to twelve, bright golden-yellow (fading to white), 2–3 cm across.

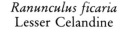

flower

> *Ranunculus ficaria*
> Lesser Celandine

11 (8) **a** Plant **with soft prickles** on leaves and stems. Basal leaves 20–40 cm long, 2–4 cm wide, elliptical-lanceolate, long-stalked, lowest may be lobed, clasping stem. Stem leaves unstalked, heart-shaped bases, clasping stem. All leaves green, hairless above, thickly white-felted below. Stem grooved, 45–120 cm tall. Flower-heads 'thistle-like', solitary or two to three together. Florets red–purple. Bracts oval, sharp-pointed, purple-tipped. Locally common in northern Britain.

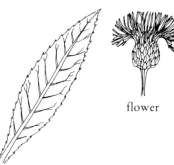

flower

> *Cirsium helenioides*
> Melancholy Thistle

b Plant **without prickles**. ⟶ 12

12 (11) **a** Plant **with square** or **angled stems**. ⟶ 13
b Plant **without square** or **angled stems**. ⟶ 15

13 (12) **a** Stem leaves **opposite**. Flowers rather 'snapdragon-like'. ——→ **14**

b Stem leaves **alternate**, oval, pointed, short-stalked. Basal leaves oval or triangular, with heart-shaped bases, blades up to 10 cm long, long-stalked. **All leaves coarsely, sharply toothed.** Plant robust, bristly-hairy, 50–100 cm tall. Inflorescence a hairy panicle. Flowers in groups of one to four on the branches. Corollas bell-shaped, 30–40 mm long, pale purplish-blue. More common in southern England.

flower
(front view)

Campanula trachelium
Nettle-leaved Bellflower

14 (13) **a** Basal leaves oblong **with heart-shaped bases**, coarsely toothed and blunt. 30–80 mm long, long-stalked. Stem leaves few, distant pairs, similar in shape to basal leaves. Plant sparsely hairy, 10–60 cm tall. Inflorescence short, oblong, whorled spike. Corollas **red–purple**. Open woods.

corolla
(front view)

Stachys (Betonica) officinalis
Betony

b Basal leaves 40–70 mm long, oblong, **shiny broadest above middle,** tapered into long stalk. Stem leaves in few pairs, shorter than basal leaves, unstalked above. Plant 10–30 cm tall with short rhizome and long leafy aerial stolons. Inflorescence spike of whorls of flowers in leaf-axils. Flower calyx bell-shaped, corolla **blue** (rarely **pink** or **white**), lower lip with white streaks.

corolla

Ajuga reptans
Bugle

15 (12) **a** Leaves **with stipules at leaf-bases**, oval, heart-shaped, wavy-edged, 5–40 mm long, long-stalked. Stems more or less hairless, 2–20 cm tall. Stipules lanceolate with wavy fringes. Flowers **'violet-like'**, petals blue–violet but variable, spur **curled up**, paler. Sepals with square-cut appendages.

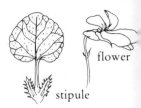

flower

stipule

> *Viola riviniana*
> Common Dog Violet

Viola reichenbachiana (Early Dog Violet) is similar to the above but leaves and stipules are **narrower**. Flowers lilac, spur **straight, pointed**, violet in colour.

b Leaves **without stipules**. Flowers **not** 'violet-like'. ⟶ **16**

16 (15) **a** Stem leaves **opposite**. ⟶ **17**
b Stem leaves **alternate**. ⟶ **18**

17 (16) **a** Basal leaves broadest above middle, narrowed into **long winged stalk**. Stem leaves **broadest below middle**, oblong, sessile or short-stalked. **All hairy**. Plant with slender creeping rhizomes with numerous prostrate non-flowering stems and erect flowering stems, 30–90 cm tall, softly hairy. Flowers numerous, 18–25 mm across, five-petalled, bright rose, tips notched. Sexes separate.

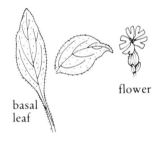

flower

basal leaf

> *Silene dioica*
> Red Campion

b Basal leaves 5–30 cm long, lanceolate to narrowly lanceolate, broadest above middle, margins entire, narrowing into a **short stalk**. Stem leaves few, **narrower than basal leaves. Sparsely hairy**. Plant 15–100 cm tall, erect or ascending. Flower-heads 1.5–2.5 cm across. Florets with mauve to dark purplish-blue corollas and four near equal lobes. Leafy bracts among florets. Damp woods.

basal leaf

inflorescence

> *Succisa pratensis*
> Devil's-bit Scabious

18 **(16) a** Basal leaves **kidney-shaped,** wavy or distant toothed, pale green, long-stalked. Stem leaves **triangular-oval with heart-shaped bases,** deeply, irregularly wavy-toothed, pale green, smelling of garlic. Plant more or less hairless, erect, 20–120 cm tall. Flowers in racemes, white, four-petalled, 6 mm across, twice as long as sepals. Open woods and wood margins.

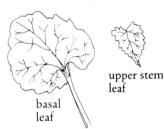

upper stem leaf

basal leaf

> *Alliaria petiolata*
> Jack by the Hedge

b Plant **without the combined features of the above.** ⟶ **19**

19 **(18) a** Leaf-stalks **winged.** ⟶ **20**
   b Leaf-stalks **not winged.** ⟶ **21**

20 **(19) a** Basal leaves **lanceolate-oblong,** broadest above middle, 15–45 cm long, narrowed into winged stalks. Stem leaves similar but upper almost sessile. Plant stout, up to 2 m tall, usually unbranched, white woolly. Flowers in a terminal raceme. Corollas up to 2 cm across, usually yellow with purple spots at bases of lobes, stamen filaments with purple hairs. Open woods.

inflorescence

> *Verbascum thapsus*
> Great Mullein

b Basal leaves **oval-lanceolate, bluntly toothed,** green, softly downy, 15–30 cm long. Stem leaves similar, shorter. Plant unbranched, 50–150 cm tall. Flowers in long, erect raceme. Corollas 4–5 cm long, tubular to narrow bell-shaped, pink–purple with dark purple spots on a white ground inside lower part of tube. Acid soils.

part of inflorescence

> *Digitalis purpurea*
> Foxglove

21 (19) **a** Basal leaves **lanceolate** or **oblong-lanceolate**. ————▶ **22**

b Leaves **broadly oval**, longer than broad, up to 40 cm long, bases heart-shaped. Leaf-stalks hollow. Plant 50–250 cm tall, **stems furrowed, often reddish**, more or less woolly. Flower-heads 'thistle-like', in racemes. Florets red–purple.

flower

> *Arctium minus*
> Lesser Burdock

basal leaf

22 (21) **a** Basal leaves lanceolate, stalked, 20–80 mm long. Stem leaves stalkless, **become progressively shorter and narrower upwards**. Plant 8–60 cm tall with short leafy runners. Flowers closed 'daisy-like'. Inflorescence a long spike, half or more the height of the plant. Florets pale brown. In dry woods.

> *Gnaphalium sylvaticum*
> Heath Cudweed

flower-head

b Basal leaves oblong-lanceolate, broadest above middle, more or less hairless, 2–10 cm long, short-stalked. Stem leaves **narrower pointed, weak-toothed**. Plant little branched. Stems leafy, 5–70 cm tall. Flower-heads 'daisy-like', small, 6–10 mm across. Ray and disc florets yellow. In dry woods, rare in southeastern Britain, avoids calcareous soils.

> *Solidago virgaurea*
> Goldenrod

Section E2

1 **a** Plant **with upper leaves** either **alternate** or **in apparent whorls of three**. Two to four distinct raised lines running along stems. Flowers four-petalled, white or some shade of pink. ─────────────▶ **2**

b Plant with **all leaves in opposite pairs** (a single pair of large opposite leaves only may be present on the flower-stem). ─────────▶ **4**

2 (1) **a** **Four distinctly raised lines** running along stem(s). ─────▶ **3**

b Stem(s) **with two distinct and two indistinct lines**. Leaves elliptical with stalks 3–20 mm long, hairless or hairy on veins, margins finely or sharply toothed. Plant 25–60 cm or more tall with many gland-tipped hairs on upper parts. Flowers at first white, later streaked pink, 4–6 mm across, stigma club-shaped. In damp woods.

flower

| *Epilobium roseum* |
| Pale Willowherb |

3 (2) **a** Stem **with numerous, slender, spreading glandular hairs**. Leaves **oval-lanceolate**, short-stalked, gradually tapering to pointed tip, **teeth small**, forwardly directed. Plant 30–90 cm or more tall, stiffly erect, much-branched. Flowers **pale pink**, 4–6 mm across, numerous, in terminal raceme, stigmas club-shaped. In damp woods in southeastern England.

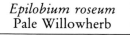

flower

| *Epilobium ciliatum* |
| American Willowherb |

b Stem **without glandular hairs**. Leaves **strap-shaped** to **narrowly oblong-lanceolate**, blunt-tipped, narrowing into more or less sessile bases, margins **strongly, irregularly toothed**. Plant erect, 25–60 cm or more tall, firm. Flowers **pale pink**, 4–6 mm across, numerous, in terminal raceme. In damp woods.

flower

| *Epilobium tetragonum* |
| Square-stemmed Willowherb |

Epilobium obscurum (Dull-leaved Willowherb) is similar but leaves are **broader lanceolate** with **distant, small teeth**. Flowers **deep rose** in colour.

4 (1) **a** Plant typically **prostrate** (stem(s) creeping along ground) and ascending at tips. ────────────────► **5**
    b Plant typically **of erect habit.** ────────────► **14**

5 (4) **a** Stems **four-angled** or **square.** ──────────► **6**
    b Stems **not four-angled** or **square.** ──────► **7**

6 (5) **a** Leaves **'ivy-like', kidney or oval heart-shaped,** 1–3 cm wide, toothed, long-stalked. Stem(s) **square, hairy.** Plant 10–30 cm long, creeping and rooting. Flowers 'snapdragon-like', two to four in a whorl, in leaf-axils. Corollas pale violet (rarely pink).

> flower
> (side view)

Glechoma hederacea
Ground Ivy

b Leaves **40–80 mm long, rigid, narrow-lanceolate** more or less hairy, rough-edged, not toothed. Stems 15–60 cm long, **roughly four-angled, weak** and **brittle.** Flowers five-petalled, 2–3 cm across, white, long-stalked, petals forked to **about** half-way. Common except on very acid soils.

> flower

Stellaria holostea
Greater Stitchwort

part of plant

Stellaria graminea (Lesser Stitchwort) is similar but leaves **no more than 40 mm long.** Petals forked to **more than** half-way. In open woods.

7 (5) **a** Leaves **with margins scallop-edged** or **more or less toothed.** Stems hairy. Corollas of flowers four-lobed, **usually some shade of blue.**
──► **8**

    b Leaves **with entire margins.** Stems may or may not be hairy. Flowers **not blue.** ──────────────────────► **11**

8 (7) **a** Leaves 20–30 mm long, oval, **coarsely toothed,** hairy on both sides, light green, stalk 5–15 mm long. Stems hairy all round, 20–40 cm long, creeping and rooting. Flowers in two- to five-flowered, loose racemes. Corollas up to 7 mm across, lilac–blue. On less acid soils.

flower

Veronica montana
Wood Speedwell

    b Leaves **at most scallop-edged or shallow-toothed.** ──────► **9**

9 (8) **a** Leaves 10–25 mm long, **oval-triangular with heart-shaped bases**, dull green, hairy. Stems **with long, white hairs on two opposite sides**. Plant 20–40 cm long. Flowers in ten- to twenty-flowered, loose racemes. Corollas up to 10 mm across, bright blue with white eye.

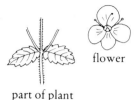

flower

Veronica chamaedrys
Germander Speedwell

part of plant

b Leaves **broadest below middle, oblong** to **elliptical-oblong**. Stems **hairy all round**. ⟶ 10

10 (9) **a** Leaves 20–30 mm long, elliptical-oblong to oval, margins scalloped or very shallowly toothed, **hairy on both sides**. Plant 10–40 cm long, creeping and rooting, forming large mats. Flowers in fifteen- to twentyfive-flowered pyramidal racemes. Corollas lilac, up to 6 mm across. In open woods.

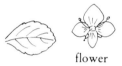

flower

Veronica officinalis
Common Speedwell

b Leaves 10–20 mm long, oval or oblong, rounded at both ends, margins almost entire, **hairless**, light green. Plant 10–30 cm long, finely downy, creeping and rooting. Flowers in up to thirty-flowered, loose, terminal raceme. Corollas white or pale blue, anthers slaty blue. In open woods.

flower

Veronica serpyllifolia
Thyme-leaved Speedwell

11 (7) **a** Plant **hairless**. Flowers yellow. ⟶ 12
    b Plant **at least slightly downy**. Flowers white. ⟶ 13

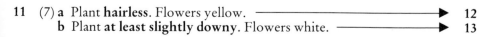
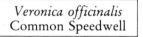

12 (11) **a** Leaves 15–30 mm long, oval to almost round **with gland dots on the surface**. Plant up to 60 cm long, creeping and rooting. Flowers solitary, cup-shaped, 15–25 mm across, on stout stalks which are shorter than the leaves. Corollas five-lobed, yellow, lobes **fringed with tiny hairs**. In moist woods.

> **Lysimachia nummularia*
> Creeping Jenny

b Leaves 20–40 mm long, oval, rounded at bases, more pointed than *Lysimachia nummularia* and **without obvious gland dots**, short-stalked. Plant up to 40 cm long. Flowers solitary, cup-shaped, five-lobed, yellow, **lobes not fringed with hairs**.

> **Lysimachia nemorum*
> Yellow Pimpernel

13 (11) **a** Leaves 3–20 mm long, oval, pointed, lower stalked, upper almost sessile. Plant 5–40 cm long, stems **with a single line of hairs**. Flowers numerous, **five-petalled**, petals **deeply two-cleft, petals not exceeding the sepals**.

> *Stellaria media*
> Common Chickweed

stem at node   flower

Stellaria nemorum (Wood Chickweed) has leaves up to 7 cm long with heart-shaped bases, all thin and sparingly hairy. **Petals twice as long as the sepals**.

b Leaves 3–20 mm long, oval to broadly elliptical **with three strong veins**, slightly downy. Flowers **four-petalled**, petals **not cleft. Sepals narrow and longer than petals**. On richer soils.

> *Moehringia trinervia*
> Three-veined Sandwort

flower

*There may be some confusion between *Veronica serpyllifolia* and *Lysimachia* species if flowers are not present. The stems of *Veronica serpyllifolia* are hairy whilst those of *Lysimachia* species are not.

14 (4) **a** Leaves **net-veined.** ⟶ **15**

    b Leaves **parallel-veined**, broad-elliptical, 5–20 cm long, **in one pair in the middle of the flower-stem.** Tiny leaves may be present on the stem above. Plant 20–60 cm tall. Flowers green–yellow, 'orchid-like', many in a loose raceme.

> *Listera ovata*
> Common Twayblade Orchid

flower (front view)

15 (14) **a** Stems **four-angled** or **distinctly square** in cross section. Flowers 'snapdragon-like'. ⟶ **16**

    b Stems **not four-angled** or **square.** ⟶ **22**

16 (15) **a** Leaves **toothed.** ⟶ **17**

    b Leaves 20–50 mm long, **not toothed** but may be wavy-edged, oval, bases rounded or wedge-shaped, stalked. Plant erect, up to 30 cm tall with creeping runners, sparingly hairy. Flowers in dense oblong head, bracts hairy, more or less purple. Corollas 10–14 mm across, violet or pink (rarely white), upper lip very concave.

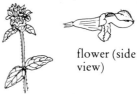

> *Prunella vulgaris*
> Self-Heal

flower (side view)

17 (16) **a** Leaves 30–70 mm long, oval, **very wrinkled surface**, base heart-shaped, blunt toothed, pointed. Plant 15–30 cm tall, hairy, with creeping rhizome and erect branched stems. Flowers in opposite pairs on loosely branched, leafless spikes. Corollas green–yellow, stamens red.

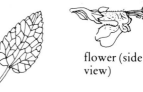

> *Teucrium scorodonia*
> Wood Sage

flower (side view)

    b Leaves oval in outline, **surface not very wrinkled.** ⟶ **18**

18 (17) **a** Plant **at least sparingly hairy**. ————————————➤ **19**

b Plant **hairless except for sticky hairs on inflorescences**, 40–80 cm tall with a short rhizome. Leaves 6–13 cm long, pointed, coarsely toothed. Inflorescence a panicle of cymes in axils of leaf-bracts. Flowers tubular. Corollas with two red–brown upper lobes and three green lower lobes.

| *Scrophularia nodosa* |
| Common Figwort |

part of inflorescence

19 (18) **a** Stem 10–100 cm tall, **bristly with red- and yellow-tipped hairs**. Leaves 2.5–10 cm long, oval, pointed, coarsely toothed, hairy, narrowing to stalk. Corollas pink or white with purple markings on lower lip.

| *Galeopsis tetrahit* |
| Common Hemp-nettle |

flower (side view)

b Stem **without red- and yellow-tipped glandular hairs**. ————➤ **20**

20 (19) **a** Plant **with stinging hairs on stems and leaves**, 30–150 cm tall with yellow, fleshy far-creeping rhizomes. Leaves 40–80 mm long, oval, heart-shaped base, pointed, coarsely toothed. Male and female flowers separate. Flowers green, without petals in drooping, catkin-like inflorescences.

| *Urtica dioica* |
| Stinging Nettle |

part of male inflorescence

b Plant **without stinging hairs**. ————————————➤ **21**

21 (20) **a** Leaves 40–90 mm long, numerous, oval with heart-shaped base, toothed. Leaf-stalks 15–70 mm long. Plant 30–100 cm tall with erect, **bristly stems** and **a creeping rhizome**. Inflorescence a loose terminal spike. Flowers claret with a white pattern on the lip.

| *Stachys sylvatica* |
| Hedge Woundwort |

flower (side view)

part of inflorescence

b Leaves 40–70 mm long, oval, pointed, rounded at base, coarsely toothed. Plant 20–60 cm tall, erect, with **long, creeping, leafy runners**. Flowers in dense axillary whorls, corollas up to 20 mm long, bright yellow with red–brown streaks.

> *Galeobdolon luteum*
> Yellow Archangel

22 (15) **a** Leaves **with margins entire.** ⟶ 23
   b Leaves **with margins wavy-edged** and/or **toothed.** ⟶ 25

23 (22) **a** Leaves **with translucent dots on the surface** (examine with lens).
   ⟶ 24

b Leaves **without translucent dots on the surface**, 2–10 cm long, oval-lanceolate to linear-lanceolate, more or less sessile. Plant 8–60 cm tall, with spreading branches. Flowers in pairs in axils of distant leaf-like bracts, rather 'snapdragon-like', deep yellow to whitish, sometimes tinged with red or purple.

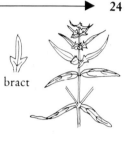

bract

> *Melampyrum pratense*
> Common Cow Wheat

flower
(side view)

24 (23) **a** Leaves 5–10 mm long **with translucent dots on the surface**, broadly oval with heart-shaped base, **half-clasping stem**. Plant hairless, 30–60 cm tall. Stem **often reddish**. Flowers in branched cymes, five-petalled, up to 15 mm across, orange–yellow, red-dotted, black dots on the edge of the petals and sepals. In dry woods.

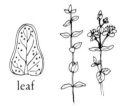

leaf

> *Hypericum pulchrum*
> Slender St John's Wort

b Leaves 10–20 mm long, elliptical to oblong, blunt, **with many translucent dots on the surface**. Plant 30–90 cm tall. Stem **with two opposite raised lines**. Flowers five-petalled, golden–yellow with black dots on the petal edges.

leaf

> *Hypericum perforatum*
> Common St John's Wort

25 (22) **a** Leaves **sharply toothed.** ━━━━━━━━━━━━▶ **26**
      b Leaves **shallowly toothed.** ━━━━━━━━━▶ **27**

26 (25) **a** Leaves 6–12 mm long, **oval** to **lanceolate.** Plant 2–30 cm tall. Stem **wiry.** Flowers 'snapdragon-like' with white corollas, lower lip longer than upper lip, upper lip often purplish. In open woods.

corolla
(front view)

> *Euphrasia officinalis aggregate*
> Common Eyebright

b Leaves **broadly lanceolate,** rounded at base, sharply, irregularly toothed, with short narrowly winged stalks. Stems **slender,** 20–60 cm tall. Flowers pale rose, four notched petals, in terminal raceme, stigmas four-lobed. In woods on base-rich soils.

flower

> *Epilobium montanum*
> Broad-leaved Willowherb

27 (25) **a** Leaves 4–10 cm long, **oval, rounded at base, tapering to tip, teeth very small** and **shallow.** Plant 20–70 cm tall, erect, with far-creeping rhizomes, very sparsely downy. Inflorescence spike-like, elongated, held well above leaves. Flowers each with two deeply cleft, white petals. In shady woods.

flower
(side
view)

fruit

> *Circaea lutetiana*
> Enchanter's Nightshade

b Leaves 3–8 cm long, **oval-elliptical, teeth small.** Stalks 3–10 mm long. Plant 15–40 cm tall. Stems unbranched, hairy. Sexes separate. Male flowers in erect, catkin-like spikes in leaf-axils. Female flowers in groups of one to three on stalks 3 cm long. In woods on basic or calcareous soils.

> *Mercurialis perennis*
> Dog's Mercury

You may identify other herbaceous plants with opposite leaves in Section E1 on page 34 beginning at **12.**

Section E3

1   **a** Leaves **in a whorl** or **rosette.** ────────▶ 2

      b Leaves **alternate.** ──────────────────▶ 3

2   (1) **a** **Parasol-like whorl** of usually four leaves at top of stem, leaves usually **diamond-shaped**, broadest above middle, pointed, 6–12 cm long. Stems erect, 15–40 cm tall. Flower solitary, long-stalked, in centre of leaf-whorl, usually with four green, lanceolate sepals and four narrower green petals.

> *Paris quadrifolia*
> Herb Paris

      b Leaves **elliptical-lanceolate, six to eight in a whorl**, pointed, **with forward-directed prickles along margins**, whorls of leaves distant. Plant erect, 15–45 cm tall, unbranched, hairy below leaf-whorls. Stems four-angled. Vanilla-scented when bruised. Flowers small, in umbel-like heads. Corollas white, lobes to half-way. On calcareous or richer soils.

> *Galium odoratum*
> Woodruff

3   (1) **a** Plant **with tubular wavy-edged stipules** on stems at leaf-bases. Leaves **oval-lanceolate**, rounded at bases, usually green and thin, occasionally rusty-red. Plant up to 100 cm tall with branches at an acute angle (20–40°). Inflorescence much-branched, leafy only at the base. Flowers in distant whorls. Sepals of flowers oblong, untoothed, blunt, up to 3 mm long, one only with a red wart.

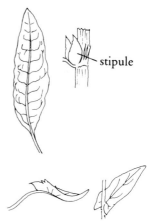

stipule

> *Rumex sanguineus*
> Wood Dock

Rumex acetosa (Sorrel) has distinctly shiny, clasping, **arrow-shaped** leaves. **Brown, fringed stipules**.

      b Plant **without stipules.** Leaves **net-veined.** ────────▶ 4

4 (3) **a** Leaves **narrow-lanceolate, wavy-edged,**
may have small distant teeth. Plant erect,
30–120 cm tall, more or less downy above.
Flowers four-petalled, rose–purple, 2–3 cm
across, borne on spikes, upper petals broader
than lower petals, stigmas four-lobed.

flower

stigma

| *Chamerion angustifolium* |
| Rosebay Willowherb |

b Leaves **of other shape, margins entire, not wavy-edged.** ⟶ **5**

5 (4) **a** Lower leaves **oval, paddle-shaped,** upper
oblong and downy. Plant 15–45 cm tall, stems
with spreading hairs. Flowers with wheel-
shaped, five-lobed corollas, 6–10 mm across,
pale blue with yellow eye. Hairs on calyx
stiffly curled or hooked. In damp woods
especially in eastern Britain.

part of
inflorescence

| *Myosotis sylvatica* |
| Wood Forget-me-not |

b Similar to *Myosotis sylvatica* but lower
leaves **broad oval,** stalked, and upper leaves
oblong-lanceolate, sessile, downy. Plant
15–30 cm tall, stems hairy. Flowers similar to
above but corollas bright blue, up to 5 mm
across, with concave lobes. In dry woods.

| *Myosotis arvensis* |
| Common Forget-me-not |

You may identify other herbaceous plants with alternate stem leaves in Section E1 on page 34 beginning at **12.**

SECTION F

1   Leaves either **single three-foliate**, or **all** or **only
some** of the leaves **twice three-foliate**.

       ━━━━━━━━━━━━━━━━━━━━━━━━━━━▶   Section F1, page 50

2   Leaves **palmately divided** or leaflets **palmately
arranged** (of five or more lobes or leaflets) arising
from the tip of the leaf-stalk.

       ━━━━━━━━━━━━━━━━━━━━━━━━━━━▶   Section F2, page 52

3   Leaves **pinnately lobed** or **cut** but not to midrib
of leaf (pinnatifid), or leaves **divided into
pinnately arranged leaflets**.

       ━━━━━━━━━━━━━━━━━━━━━━━━━━━▶   Section F3, page 53

Section F1

1   **a** Some leaves **twice three-foliate.** ⟶ 2
     b Leaves **typically single three-foliate.** ⟶ 3

2   (1) **a** Stem leaves **usually single three-foliate,** leaves triangular in outline, leaflets oval or elliptical, 20–70 mm long, irregularly toothed, fresh light green. Leaf stalks **triangular in section.** Plant erect, hairless, 40–100 cm tall. Stems stout, hollow, grooved. Flowers small, five-petalled, **white,** in many-rayed terminal umbels, 20–60 mm across.

> *Aegopodium podagraria*
> Ground Elder

inflorescence

b Leaves hairless, leaflets deeply lobed, **lobes tipped with tiny spines,** each leaflet of the basal leaves may be further cut into three leaflets, leaves long-stalked, **two opposite, single three-foliate leaves** on flower-stem, all leaves fleshy, grey–green. Plant up to 12 cm tall with a creeping, scaly rhizome. Inflorescence a long-stalked head of five, **pale yellow–green** flowers. Four flowers face outwards at right-angles and one faces upwards.

> *Adoxa moschatellina*
> Townhall Clock

3   (1) **a** Leaves **divided into three separate leaflets** (leaflets may be further deeply divided). ⟶ 4
     b Basal root leaves **three-lobed** (some may be **kidney**-shaped), lobes coarsely toothed. Stem leaves few, **deeply divided into lanceolate segments.** Stems numerous, erect, sparsely hairy. Plant 10–40 cm tall. Flowers 'buttercup-like', few, up to 25 mm across, five yellow petals. On basic soils.

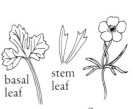

basal leaf   stem leaf

flower

> *Ranunculus auricomus*
> Goldilocks

4 (3) **a** Leaves mainly basal, on stalks up to 15 cm long, leaflets **with entire margins, drooping, broadest above middle** and **notched at tips,** yellow–green above, purplish below, 10–20 mm long. Plant creeping. Flowers on long stalks, solitary, five-petalled, white with violet veins. In dry woods, especially oak and beech woods.

> *Oxalis acetosella*
> Wood Sorrel

b Leaflet **margins toothed** or **cut into lobes.** ⟶ **5**

5 (4) **a** Leaf margins **cut into lobes.** ⟶ **6**
b Leaf margins **toothed**, leaflets 10–60 mm long, oblong, terminal tooth **longer than** side teeth. Basal leaves long-stalked. All leaves bright, glossy green. Plant 5–30 cm tall with **long arching runners**. Flowers with five white petals, each 12–18 mm across. Fruits tiny 'strawberries'.

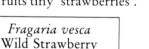

> *Fragaria vesca*
> Wild Strawberry

Potentilla sterilis (Barren Strawberry) has **rounded leaflets with five to seven teeth on each side**, terminal tooth **shorter than** side teeth. Plant with **short runners**.

6 (5) **a** Leaves **hairy**, leaflets three- to five-lobed (**palmate**). Stem leaves **alternate**, sessile with narrow segments. Stems hairy, furrowed, 15–60 cm tall. Plant with runners which root at nodes. Flowers 'buttercup-like', up to 30 mm across, of five **deep yellow** petals.

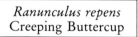

> *Ranunculus repens*
> Creeping Buttercup

stem
leaf   flower

b Leaves **hairless**, leaflets thin, dark green, **deeply lobed**, segments narrow-oblong, sharply toothed. Few, long-stalked **basal leaves arising from a slender rhizome**. Stem leaves three, stalked, **arranged in a whorl** below flower. Flower-stems 6–30 cm tall, sparsely hairy. Flowers solitary, 20–40 mm across, 'buttercup-like' with five **white pink-tinged**, petal-like sepals.

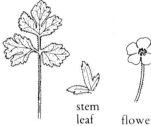

flower

> *Anemone nemorosa*
> Wood Anemone

Section F2

1   **a** Leaves divided into **distinct triangular leaflets**, leaflets further deeply pinnately cut. Lower leaves long-stalked. Upper leaves alternate, **divided into more or less equal triangular parts**. All leaves shiny green. Plant 10–50 cm tall. Stems usually branched from base, often red-tinged, brittle, sparsely hairy. Flowers with five, bright pink petals (sometimes white), anthers orange or purple.

> *Geranium robertianum*
> Herb Robert

    b Leaves and/or leaflets **palmately lobed.** ⟶ **2**

2 (1) **a** Leaf outline **palmate**. Basal root leaves **palmately veined**, variable in outline. Stem leaves divided into narrow lobes.

> See *Ranunculus auricomus*, page 50

    b Leaves typically **three-foliate with palmate lobes.** ⟶ **3**
    c Leaves typically **five or more palmately lobed.** ⟶ **4**

3 (2) **a** Leaves hairless, lobes **tipped with a tiny spine**.

> See *Adoxa moschatellina*, page 50

    b Leaves hairy, leaflets further deeply cut but lobes **without a tiny spine at the tip.**

> See *Anemone nemorosa*, page 51

4 (2) **a** Leaves **opposite**, up to 15 mm long, shallowly lobed, **like 'ivy leaves'**, two to three large teeth or small lobes on each side, light green. Plant hairy and spreading, 10–60 cm long. Flowers solitary, in leaf-axils, corollas four-lobed, pale lilac.

flower

> *Veronica hederifolia*
> Ivy-leaved Speedwell

    b Leaves **more deeply cut.** ⟶ **5**

5 (4) **a** Leaves **cut but not into distinct leaflets.** ————————➤ 6

b Leaves appearing as **five toothed leaflets** (three leaflets and two lower, shorter stipules), 5–10 mm long with coarse teeth, shiny deep green, silky below. Basal leaves long-stalked. Stems creeping to erect, up to 10 cm tall. Flowers many, yellow, four-petalled, in loose terminal cymes.

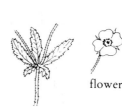

flower

> *Potentilla erecta*
> Tormentil

6 (5) **a** Lobes pointed, **cut to three-quarters of the base,** lobes further cut and toothed. Root leaves long-stalked, in basal rosette. Stem leaves alternate. Plant hairless, 20–60 cm tall. Flowers small, oval, five-petalled, white or pink, in small, few-rayed, irregular umbels. On richer soils, especially in beech woods.

> *Sanicula europaea*
> Sanicle

b Basal root leaves **five to seven, deeply palmately lobed, cut nearly to base.** Lobes elliptical broadest above middle, further cut. Stem leaves alternate, similar to root leaves but two basal lobes may be separate from rest of leaf. Leaves pale green. Plant erect and hairy, upper parts with sticky hairs. Flowers with five, pinkish-violet or reddish corollas. Damp woods, especially in northern England and Scotland.

flower

> *Geranium sylvaticum*
> Wood Crane's-bill

Section F3

1   **a** Leaves **pinnately cut** (pinnatifid). ————————➤ 2

   b Leaves **divided into pinnately arranged leaflets** (may be further divided). ————————➤ 5

2 (1) **a** Plant 'thistle-like' **bearing spines** on leaves and stems. ————➤ 3

   b Plant not 'thistle-like', **without spines.** ————————➤ 4

3 (2) **a** Basal leaves lanceolate in outline, broadest above middle, deeply cut and wavy-edged, tipped with strong spines. Stem leaves sessile, alternate. All leaves hairy and **prickly**, cottony below. Plant 30–150 cm tall. Stem furrowed, **with interrupted spiny wings**. Flower-heads solitary or two to three in a cluster. Florets pale red–purple. Outer bracts with long, arched-back, yellow-spined tips. In open woods.

flower-head

stem leaf

> *Cirsium vulgare*
> Spear Thistle

b Basal root leaves lanceolate, less deeply lobed than upper leaves, margins spiny. Stem leaves sessile, alternate, deeply cut and wavy. All leaves hairy and slightly cottony below, dark green and often purple-flushed. Plant 30–150 cm tall. Stem furrowed, **with continuous spiny wings, hairy and cottony**. Flower-heads in crowded clusters. Florets dark purple (sometimes white). Outer bracts purplish. In open woods.

flower-heads

> *Cirsium palustre*
> Marsh Thistle

4 (2) **a** Lower leaves **oblong-lanceolate** in outline, short-stalked lobes **unequal** and **more or less further cut**. Upper leaves alternate, oblong, sessile. All leaves cottony at first, later hairless, yellow–green. Plant erect, 30–70 cm tall. Stems slender, grooved, somewhat cottony. Flowers small, in flat-topped corymbs. Ray florets yellow. Outer bracts small, inner bracts sticky-hairy. In open woods.

> *Senecio sylvaticus*
> Wood Groundsel

b Lower leaves **lyre-shaped**, lobes **triangular**, end lobes larger, three-lobed. Upper leaves alternate, stalkless, smaller and less lobed than lower leaves, bases arrow-shaped. Leaves thin and may be red-tinged. Plant erect, hairless, 25–100 cm tall. Flower-heads small, flowers with five yellow petals, on an open panicle branched at 90°. On calcareous soils, especially in beech woods.

> *Mycelis muralis*
> Wall Lettuce

5 (1) a Leaves **one-pinnate**. ⟶ **6**

b Main **pinnate leaflets further divided into smaller leaflets which may be further cut.** ⟶ **18**

6 (5) a At least lower leaves with side leaflets **of a different size** (in some, large leaflets alternate regularly with smaller leaflets). Stipules present at base. ⟶ **7**

b Side leaflets **not obviously of a different size.** Stipules may or may not be present at leaf-bases. ⟶ **10**

7 (6) a Terminal leaflet **very large compared with side leaflets, more or less triangular** and **three-lobed**. Two to three pairs of unequal side leaflets, 5–10 mm long. Stipules large and leafy. Upper leaves three-foliate or undivided. Plant downy, up to 60 cm tall. Flowers up to 15 mm across with five spreading yellow, rounded petals, on long erect stalks. On less acid soils.

lower leaf
Geum urbanum

> *Geum urbanum*
> Herb Bennet

Geum rivale (Water Avens) has leaves with **a rounded** or **kidney-shaped, unlobed** terminal leaflet with small, sharp-teeth. Flowers drooping with erect orange–pink petals.

lower leaf
Geum rivale

b Terminal leaflet **not obviously larger than side leaflets.** Side leaflets may increase in size upwards. ⟶ **8**

8 (7) **a** Basal leaves 30–60 cm long, **arising from a rhizome**, with **two to five pairs of similarly sized main leaflets**, 20–80 mm long, oval, sharply double-toothed, dark green, pale green or white woolly below. Upper leaves alternate. Stipules rounded, leafy. Plant 60–100 cm tall. Flowers in dense irregular umbel-like inflorescences. Flowers five-petalled, creamy, fragrant, many stamens.

> *Filipendula ulmaria*
> Meadowsweet

b Side leaflets becoming **larger towards the top of the leaf.** ——▶ 9

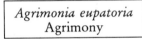

9 (8) **a** Largest leaflets up to 6 cm long, strongly toothed. Plant up to 60 cm tall. Stems **often reddish**. Flowers many, with five petals, on long spikes. Calcareous soils.

> *Agrimonia eupatoria*
> Agrimony

part of inflorescence

b Similar to *Agrimonia eupatoria* but more robust and up to 1 m tall. Stems **never reddish**. Leaves larger and leaflets larger and relatively narrower, more pointed with conspicuous shining yellow glands on under side. Flowers six to eight.

> *Agrimonia procera*
> Fragrant Agrimony

10 (6) **a** Leaf-base **with stipules.** ———————▶ 11
   b Leaf-bases **without stipules.** —————▶ 14

11 (10) **a** Side leaflets **one to three pairs per leaf**. Stems **winged.** ——▶ 12
   b Side leaflets **five to twelve pairs of leaflets**. Stems **round.** ——▶ 13

12 (11) **a** One pair of side leaflets, narrowly lanceolate. Tip of leaf **extending into a branched tendril.** Stipules up to 20 mm long. Stems broadly winged. Plant up to 3 m tall, climbing.

> See *Lathyrus sylvestris,* page 9

b Two to four pairs of side leaflets, narrow lanceolate, pointed, up to 4 cm long. **Tendrils absent.** Stipules lanceolate. Plant 15–40 cm tall. Flowers 'pea-like' in two- to six-flowered racemes. Flowers crimson–red turning blue or green. In hilly woods on more acid soils.

> *Lathyrus montanus*
> Bitter Vetch

13 (11) **a** Leaflets **oval, widest near base,** five to six pairs of side leaflets. Tendrils present. Plant climbing.

> See *Vicia sepium,* page 9

b Leaflets **downy, elliptical, widest near middle,** six to twelve pairs of side leaflets. Tendrils present. Stipules half-arrow-shaped.

> See *Vicia cracca,* page 9

14 (10) **a** Pinnate basal leaves further divided but soon withering. Stem leaves **one-pinnate with narrow-linear segments.** Plant up to 50 cm or more tall. Stems smooth and slender, greatly narrowed near base. Flowers small, white, five-petalled, on six- to twelve-rayed umbels, 30–70 mm across.

> *Conopodium majus*
> Pignut

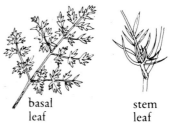

basal leaf   stem leaf

b Basal leaves (if present) may have leaflets of different shape to stem leaflets but all are **one-pinnate.** ————————————➤ **15**

15 (14) **a** Stems 50–200 cm tall, stout, **coarsely, stiffly hairy, hairs downward pointing**. Leaves 15–60 cm long, leaflets 5–15 cm long, variously divided, toothed, oval to narrowly lanceolate, lower stalked. Flowers whitish or pinkish in many-rayed umbels, 5–15 cm across.

> *Heracleum sphondylium*
> Cow Parsnip

b Stems **not coarsely, stiffly hairy**. ────────► **16**

16 (15) **a** Leaflets of lower leaves **toothed on the lower side only**, leaves opposite, leaflets lanceolate, side veins conspicuous. Basal leaves with a terminal leaflet up to 20 cm long. Plant 30–120 cm tall. Stems hairy below. Inflorescence terminal umbel-like head with five-lobed, funnel-shaped, pinkish–white corollas.

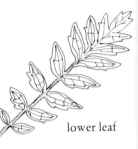

lower leaf

> *Valeriana officinalis*
> Common Valerian

b Leaflets **without teeth on the lower side**. ────────► **17**

17 (16) **a** Terminal leaflet of basal leaves **as broad as long**, rather kidney-shaped, much broader than side leaflets. Basal leaves long-stalked, leaflets narrowly lanceolate, terminal may be three-lobed. Plant erect or ascending, 30–60 cm, with runners. Flowers in racemes, four-petalled, lilac (rarely white), anthers yellow.

> *Cardamine pratensis*
> Lady's Smock

b Basal leaves with the terminal leaflet **longer than broad**. All leaflets of upper stem leaflets oval. Plant 10–15 cm tall. Stem leafy, wavy, growing from a cluster of basal leaves. Flowers with four white petals, anthers yellow, six in number.

leaf

> *Cardamine flexuosa*
> Wood Bittercress

18 (5) **a** Leaves **'fern-like'** in outline. ⟶ **19**

b Leaflets **oval**, fresh green, sharp toothed. Basal leaf-stalks channelled on upper side. Upper leaves reduced to sheaths around flower umbels. Plant up to 200 cm tall. Stems hollow, grooved, very stout, purplish. Flowers small, white, five-petalled on many-rayed terminal umbels, 3–15 cm across.

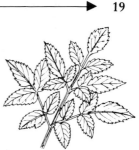

> *Angelica sylvestris*
> Wild Angelica

19 (18) **a** Leaves up to 30 cm long, fresh green, leaflets pointed, **coarsely toothed**, downy. Plant downy, erect, 60–100 cm tall. Stems **hollow**, furrowed. Flowers small, pure white, five-petalled on four- to ten-rayed umbels, up to 6 cm across. Wood margins.

> *Anthriscus sylvestris*
> Cow Parsley

b Leaves **narrower than** *Anthriscus sylvestris*, roughly hairy, dull green. Plant 5–125 cm tall. Stems **solid**, rough, hairy. Flowers small, white, pinkish or purplish, five-petalled, on five- to twelve-rayed umbels. Wood margins.

> *Torilis japonica*
> Upright Hedge Parsley

Index

Acer campestre, 18
  pseudoplatanus, 18
Adoxa moschatellina, 50, 52
Aegopodium podagraria, 50
Aesculus hippocastanum, 16
Agrimonia eupatoria, 56
  procera, 56
Agrimony, 56
  Fragrant, 56
Agropyron caninum, 28
Agrostis canina, 27
  tenuis, 27
Ajuga reptans, 35
Alder, 23
Alliaria petiolata, 37
Allium ursinum, 32
Alnus glutinosa, 23
Anemone nemorosa, 51, 52
Anemone, Wood, 51
Angelica sylvestris, 59
Angelica, Wild, 59
Anthoxanthum odoratum, 27
Anthriscus sylvestris, 59
Archangel, Yellow, 45
Arctium minus, 38
Arum maculatum, 33
Ash, 17
Aspen, 24
Athyrium filix-femina, 8
Avens, Water, 55

Bedstraw, Heath, 11
Beech, 22
Bellflower, Nettle-leaved, 35
Bent, Common, 27
  Velvet, 27
Betonica officinalis, 35
Betony, 35
Betula pendula, 21
  pubescens, 21
Bindweed, Black, 13
  Hedge, 13
Birch, Downy, 21
  Silver, 21
Bittercress, Wood, 58
Bittersweet, 11
Blackthorn, 14
Blechnum spicant, 7
Bluebell, 32
Brachypodium sylvaticum, 27
Bracken, 8
Bramble, 10
Brome, Wood False, 27
Broom, 16

Bryony, Black, 13
Buckthorn, 14
  Alder, 19
Bugle, 35
Burdock, Lesser, 38
Buttercup, Creeping, 51

Calystegia sepium, 12
Campanula trachelium, 35
Campion, Red, 36
Cardamine flexuosa, 58
  pratensis, 58
Carex flacca, 30
  hirta, 30
  paniculata, 29
  pendula, 29
  remota, 29
  sylvatica, 30
Carpinus betulus, 23
Castanea sativa, 20
Celandine, Lesser, 34
Chamerion angustifolium, 48
Cherry, Wild, 22
Chestnut, Horse, 16
  Sweet, 20
Chickweed, Common, 42
Christmas Tree, 15
Circaea lutetiana, 46
Cirsium helenioides, 34
  palustre, 54
  vulgare, 54
Cleavers, 11
Clematis vitalba, 11
Conopodium majus, 57
Convallaria majalis, 31
Corylus avellana, 24
Couch, Bearded, 28
Cow Wheat, Common, 45
Crane's-bill, Wood, 53
Crataegus laevigata, 15
  monogyna, 14
Crosswort, 11
Cudweed, Heath, 38
Cytisus scoparius, 16

Dactylorhiza fuchsii, 33
Daphne laureola, 19
Deschampsia caespitosa, 28
  flexuosa, 25
Dewberry, 10
Digitalis purpurea, 37
Dock, Wood, 47
Dogwood, 17
Dryopteris dilatata, 8

  filix-mas, 8

Elder, 17
  Ground, 50
Elm, English, 21
  Wych, 21
Endymion non-scriptus, 32
Epilobium ciliatum, 39
  montanum, 46
  obscurum, 39
  roseum, 39
  tetragonum, 39
Euonymus europaeus, 18
Euphorbia amygdaloides, 31
Euphrasia officinalis, 46
Everlasting Pea, Narrow-leaved, 9
Eyebright, Common, 46

Fagus sylvatica, 22
Fallopia convolvulus, 13
Fern, Broad Buckler, 8
  Hard, 7
  Harts Tongue, 7
  Lady, 8
  Male, 8
Fescue, Giant, 28
Festuca gigantea, 28
  rubra, 26
Figwort, Common, 44
Filipendula ulmaria, 56
Forget-me-not, Common, 48
  Wood, 48
Foxglove, 37
Fragaria vesca, 51
Frangula alnus, 19
Fraxinus excelsior, 17

Galeobdolon luteum, 45
Galeopsis tetrahit, 44
Galium aparine, 11
  cruciata, 11
  odoratum, 11, 47
  saxatile, 11
Geranium robertianum, 52
  sylvaticum, 53
Geum rivale, 55
  urbanum, 55
Glechoma hederacea, 40
Gnaphalium sylvaticum, 38
Goldenrod, 38
Goldilocks, 50
Groundsel, Wood, 54

Hair-grass, Tufted, 28
  Wavy, 26

Hawthorn, 14
  Midland, 15
Hazel, 24
Hedera helix, 9
Hemp-nettle, Common, 44
Heracleum sphondylium, 58
Herb Bennet, 55
  Paris, 47
  Robert, 52
Holcus mollis, 26
Holly, 19
Honeysuckle, 12
Hop, 12
Hornbeam, 23
Humulus lupulus, 12
Hyacinthoides non-scriptus, 32
Hypericum perforatum, 45
  pulchrum, 45

Ilex aquifolium, 19
Ivy, 9
  Ground, 40

Jack by the Hedge, 37
Jenny, Creeping, 42
Juniper, 16
Juniperus communis, 16

Larch, Common, 15
  Japanese, 15
Larix decidua, 15
  kaempferi, 15
Lathyrus montanus, 57
  sylvestris, 57
Lettuce, Wall, 55
Lily of the Valley, 31
Lime, Common, 24
  Small-leaved, 24
Listera ovata, 43
Lonicera periclymenum, 12
Lords and Ladies, 33
Luzula multiflora, 25
  pilosa, 25
  sylvatica, 25
Lysimachia nemorum, 42
  nummularia, 42

Maple, Field, 18
Meadowsweet, 56
Melampyrum pratense, 45
Melica uniflora, 26
Melick, Wood, 26
Melilot, Tall, 10
Melilotus altissima, 10
Mercurialis perennis, 46
Mercury, Dog's, 46
Moehringia trinervia, 42

Mullein, Great, 37
Mycelis muralis, 55
Myosotis arvensis, 48
  sylvatica, 48

Nettle, Stinging, 44
Nightshade, Enchanter's, 46

Oak, Pedunculate, 20
  Sessile, 20
Orchid, Common Spotted, 33
  Common Twayblade, 40
  Early Purple, 32
  Greater Butterfly, 32
Orchis mascula, 32
Oxalis acetosella, 51

Paris quadrifolia, 47
Parsley, Cow, 59
  Upright Hedge, 59
Parsnip, Cow, 58
Phyllitis scolopendrium, 7
Pignut, 57
Pimpernel, Yellow, 42
Picea abies, 15
Pine, Scots, 15
Pinus sylvestris, 15
Platanthera chlorantha, 32
Polygonum convolvulus, 13
Polypodium vulgare, 7
Polypody, Common, 7
Poplar, Grey, 24
Populus canescens, 24
  tremula, 24
Potentilla erecta, 53
  sterilis, 51
Primrose, 34
Primula vulgaris, 34
Prunella vulgaris, 43
Prunus avium, 22
  spinosa, 14
Pteridium aquilinum, 8

Quercus petraea, 20
  robur, 20

Ramsons, 32
Ranunculus auricomus, 50, 52
  ficaria, 34
  repens, 51
Rhamnus catharticus, 14
Rosa arvensis, 10
  canina, 14
Rose, Dog, 14
  Field, 10
  Guelder, 18
Rowan, 17

Rubus caesius, 10, 14
  fruticosus, 10
Rumex acetosa, 47
  sanguineus, 47

Sage, Wood, 43
Salix alba, 20
  aurita, 23
  caprea, 22
  cinerea, 23
  fragilis, 21
Sambucus nigra, 17
Sandwort, Three-veined, 42
Sanicle, 53
Sanicula europaea, 53
Scabious, Devil's-bit, 36
Scrophularia nodosa, 44
Sedge, Drooping, 29
  Glaucous, 30
  Hairy, 30
  Remote, 29
  Tussock, 29
  Wood, 30
Self-heal, 43
Senecio sylvaticus, 54
Silene dioica, 36
Sloe, 14
Smock, Lady's, 58
Soft-grass, Creeping, 26
Solanum dulcamara, 11, 12
Solidago virgaurea, 38
Sorbus aria, 22
  aucaparia, 17
Sorrel, 47
  Wood, 51
Speedwell, Common, 41
  Germander, 41
  Ivy-leaved, 52
  Thyme-leaved, 41
  Wood, 40
Spindle Tree, 18
Spruce, Norway, 15
Spurge Laurel, 19
Spurge, Wood, 31
Stachys officinalis, 35
  sylvatica, 44
Stellaria graminea, 40
  holostea, 40
  media, 42
Stitchwort, Greater, 40
  Lesser, 40
St John's Wort, Common, 45
  Slender, 45
Strawberry, Barren, 51
  Wild, 51
Succisa pratensis, 36
Sycamore, 18

Tamus communis, 13
Taxus baccata, 16
Teucrium scorodonia, 43
Thelycrania sanguinea, 17
Thistle, Marsh, 54
  Melancholy, 34
  Spear, 54
Tilia cordata, 24
  europaea, 24
Torilis japonica, 59
Tormentil, 53
Townhall, Clock, 50
Traveller's Joy, 11

Ulmus glabra, 21
  procera, 21
Urtica dioica, 44

Valeriana officinalis, 58
Valerian, Common, 58
Verbascum thapsus, 37

Vernal Grass, Sweet, 27
Veronica chamaedrys, 41
  hederifolia, 50
  montana, 40
  officinalis, 30
  serpyllifolia, 41
Vetch, Bush, 9, 57
  Tufted, 9, 57
Viburnum lantana, 18
  opulus, 18
Vicia cracca, 9, 57
  sepium, 9, 57
Viola hirta, 33
  odorata, 33
  reichenbachiana, 36
  riviniana, 36
Violet, Common Dog, 36
  Early Dog, 36
  Hairy, 33
  Sweet, 33

Wayfaring Tree, 18
Whitebeam, 22
Willow, Crack, 21
  Eared, 23
  Goat, 22
  Grey, 23
  White, 20
Willowherb, American, 39
  Broad-leaved, 46
  Dull-leaved, 39
  Pale, 39
  Rosebay, 48
  Square-stemmed, 39
Woodruff, 47
Woodrush, Great, 25
  Hairy, 25
  Many-headed, 25
Woundwort, Hedge, 44

Yew, 16

Bibliography

Bunce, R.G.H. (1982). *A Field Key for Classifying British Woodland Vegetation*, Part 1. Institute of Terrestrial Ecology, Cambridge.

Clapham, A.R., Tutin, T.G. and Warburg, E.F. (1981). *Excursion Flora of the British Isles* (3rd edn.). Cambridge University Press, Cambridge.

Fitter, R., Fitter, A. and Farrer, A. (1984). *Guide to the Grasses, Sedges, Rushes and Ferns of Britain and Northern Europe*. Collins, London.

Hubbard, C.E. (1984). *Grasses* (3rd edn.). Penguin Books, London.

Jermy, A.C., Chater, A.O. and David, R.W. (1982). *Sedges of the British Isles*. Botanical Society of the British Isles.

McClintock, D. and Fitter, R.S.R. (1956). *Pocket Guide to Wild Flowers*. Collins, London.

Mitchell, A. (1978). *A Field Guide to the Trees of Britain and Northern Europe*, Collins, London.

Phillips, R. (1978). *Trees in Britain, Europe and North America*. Book Club Associates, London.

Rose, F. (1981). *The Wild Flower Key*. Frederick Warne, London.